*Alessandro Pirrone*

# *Ecological Policy Handbook Volume I*

## *Overview and limits*

# INTRODUCTION

## *Live and Survive*

Among the most fascinating and fundamental concepts of the science of life are those of knowing how to live and survive.

Still, talking about *"way of life"* there are also those concerning sustainability and resilience to which entire research and work tables are dedicated. Official bodies globally coordinated by extraordinary and popular authors try to attract attention of the *"political world"* sharing they insight, deeply delve into theories and practices of resilience and sustainability and which encompass the two most diverse aspects that we often mention, without much reflection, forgetting their most intrinsic meaning: *how to live and survive*.

For some time now, a real scientific alliance has been created between several pleasant authors, universities and institutes, born in the second half of the 90ˢ and which has found inspiration in the work of the ecologist Crawford Holling. This synergistic meeting of knowledge was followed by the publication of the online magazine *"Ecology and society"* with the aim of gathering reflections, analysis and useful researches in the areas of resilience and sustainability.

In short, since the early seventies, he defined "resilience" the ability of natural systems or integrated ecological and human systems, to absorb a disturbance and to reorganize while change takes place, so as to maintain the same functions, the same structure, the same identity and the same feedback: in practice *"surviving"*.

Therefore, the system has the possibility of evolving into multiple states, different from the one preceding the disturbance, associated with maintaining the vitality of the functions and structures of the system itself.

Resilience, according to Holling, is measured by the degree of disturbance that can be absorbed before the system changes its structure, changing variables and processes that control its behavior. For an ecosystem, therefore, the ability to tolerate a disturbance is to recognize its collapse in a different qualitative state, which will then be controlled by a different series of processes.

The concept of resilience has also been used in a similar way in engineering and other scientific fields. In fact, Holling's theory is widely accepted, and therefore *populations* and, by inference, *ecosystems* have more than one state of equilibrium and after a perturbation they often restore a different equilibrium from the previous one.

Crawford Holling had the merit of having applied the results of the continuous analysis of integrated adaptive systems to ecology, providing to the disciplines of applied ecology, ecosystem management and the integrated vision of ecology, economics and social sciences (and therefore of the science of sustainability as a whole) - a contributions of the highest level and depth.

Resilience scholars recognize four characteristics of resilience which are defined as latitude, resistance, precariousness and panarchy.

*Latitude* is the maximum amount in which a system can change without losing its recovery ability (therefore, before crossing a "threshold" which, once passed, can make recovery difficult or impossible).

*Resistance*, on the other hand, constitutes the ease or difficulty of changing the system, or rather, how much and how the system is resistant overall to change.

*Precariousness* indicates how close the current state of a system is to a limit or a threshold.

*Panarchy* (which recalls the Greek god Pan) is a term that is used to remember that, due to interactions at different scales, the resilience of a system to a particular scale will depend on the influences of states and dynamics on the scales that take place above or below the system itself.

Let's explore the concept of "scale".

For simplification, we report a brief semantic description of the term "scale" so as to have no more doubts in continuing to read, given that the term has been repeated many times.

The scale can be and is:

- a fixed structure with steps that allows you to go up or down from one level to another in buildings or in open places; can be spiral, on a circular plan, which rises in a helix and occupies very little space; mobile; a stepped conveyor belt used for the simultaneous transport of people from one floor to another.

- it is a transportable tool to go up or down from different levels, consisting of two uprights equipped with connecting elements placed at equal distance between them called pegs.

- it is an ordered succession of concrete elements or homogeneous sizes and values, arranged according to different criteria: it can be musical, like an ordered succession of sounds that make up a musical system; of colors and their gradation; used to measure temperatures such as Celsius, Fahrenheit, Kelvin; graduate, in mathematics, like the straight line on which a gradation of indicated values is distributed; used in geology to measure the degree of intensity of earthquakes and similar phenomena such as Mercalli or Richter; the escalator, in economics, is a mechanism for automatic adjustment of wages or other to the cost of living; it is an increasing or decreasing order of height or other dimensions or characteristics; it is a form - ladder-shaped; it is a greatness - scaled; it is a geometric modality - scale arrangement.

- in cartography and in technical drawing, or in plastics, it is the relationship between the measurements of real objects and those of their representation: reproduce on a scale.
- it is a measure or a dimension: to operate on a national scale, on a wide or large scale, in large proportions, or in a very large sector or area; on a small scale, in a limited sector, to a modest extent; the optimal scale.

We are interested in the latter definition: measurement or size, a measurability.

The scale laws, therefore, are those laws that result from the study of the behavior of a material system (air, water, oxygen, $CO_2$ etc.) following a variation of the parameters that characterize its own scale, in particular in phase transitions and in those order-disorder.

For example: what effects does a temperature rise have on a gas?

- at constant pressure, it produces an increase in volume (Gay-Lussac law); at constant volume, it produces an increase in pressure (Charles's law).

The scale or *invariance*[1] properties are those that manifest themselves in the course of a physical phenomenon as the scale of the quantities that describe them change: for example in the expansion of metals.

The concepts of law of scale and invariance of scale play a central role in the analysis of the increasingly complex systems that are studied in the physical sciences, but also in many other fields. Many systems - starting from the very large-scale structure of the universe, passing through the complex forms of biological structures, to the

---

[1] In mathematics, an invariant is a property, held by a class of mathematical objects, which remains unchanged when transformations of a certain type are applied to the objects. The particular class of objects and type of transformations are usually indicated by the context in which the term is used.

elementary interactions between the fundamental constituents of matter - show well-defined laws of scale. These laws characterize the change in the properties of the system under the effect of a transformation of the length scale and represent an essential element for understanding the *"complexity"* of the system.

For a system like an atom, the laws of scale are not particularly interesting. In fact, if we consider a scale of lengths in the order of the atom, we could adequately define all its properties, as the presence of a central nucleus and the distribution of electrons around it. However, if we consider a much larger scale, the atom itself becomes in all respects point-like and does not show particularly interesting properties.

This situation completely changes if we consider a very familiar but quite complex structure like that of a tree. In this case we can start from atoms, which form molecules, which then form cells, fibers, then leaves and branches of different sizes with ramifications both at small and large scales; finally we come to the whole tree, which for much larger scales can also be considered point-like. For this system there is a wide variety of scales whose properties are quite similar: for example the branch bifurcation occurs for both small and larger branches. In this context - where we can see on the minimum scale the fibers, and on the maximum scale the trunk or the tree itself - we can define an approximate invariance of scale with its characteristic properties, which are essential for understanding the complexity of the structure and its functionality and represent one of the peculiarities that underlie the complex structures. Other familiar examples could be represented by the structure of the lungs or arteries, but there are well-defined laws of scale in very different fields, such as seismology, meteorology, economics. In general we can find these properties in all those systems consisting of a large number of elements that interact in a non-linear way and this can explain the intrinsic cross disciplinary and generality of the scale invariance properties, whose discovery has changed the way we see both natural and social phenomena.

A concept that can be considered somewhat the inverse of resilience is that of *"vulnerability"*. Vulnerability takes place when an ecological or social system loses its resilience skills thus becoming vulnerable to the change that previously could be absorbed.

In a resilient system, change has the potential to create opportunities for development, novelty, and innovation. In a vulnerable system, even small changes can be devastating. The vulnerability therefore refers to the propensity of a socio-ecological system to suffer severely from exposure to stress and external shocks. The less resilient the system, the lower the ability of institutions and companies to adapt and cope with changes.

Implementing sustainability policies therefore means learning how to manage uncertainty, adapting to the changing conditions that arise but, above all, avoid making natural systems and our social systems less and less resilient. We are in a world where humanity is playing a prominent role in changing the processes of the biosphere, from the genetic level to the global scale. We have an extreme need to mitigate our impact on natural systems and to hone our skills to adapt to new situations, with great learning skills and flexibility.

Sustainability policies based on the best trans-disciplinary scientific knowledge should become the priority of international political agenda. The environmental, economic and social cost that we could pay, if this does not take place, could in fact be very high.

We have mentioned panarchy.

In addition to what has been said, the term panarchy also indicates the definition of the other two important aspects of the dynamics of the social and ecological systems, mentioned above, adaptability and transformability.

*Adaptability*, or the ability of the actors of a system to influence the resilience of the system itself; in a social and ecological system this characteristic concerns precisely the ability to manage resilience by human intervention.

*Transformability*, that is what constitutes the ability to fundamentally create a new system when ecological, economic and social conditions make the system no longer maintainable: the ability to maintain an adaptive system and therefore operate with adaptive governance helps to create adaptability and transformability in the same social and ecological systems.

Now, whoever is interested in understanding what's going on, raise your hand.

Just put your hand down, let's turn the page.

# CONCEPTUAL FRAMEWORK

# Exploring the Scale Issues

*What is the "Scale Problem"?*

Scale problems refer to the threats posed by economic activities to global life support systems such as the atmospheric ozone layer's protection against ultraviolet radiation, and the carbon cycle's provision of climate stability. Human economic activities are now threatening these natural systems at both the local and global levels for the first time in the history of the planet.

*Why is the Scale Problem Important?*

Scale problems are pervasive and unprecedented. The life support systems addressed by the scale issue are essential and irreplaceable ecosystem services that literally make life on earth possible. If these life support services are damaged beyond repair, then human civilization as we know it will collapse. There are currently many scale related problems that are unprecedented in human history, and that will irrevocably damage these life support functions if not corrected.

*What are some examples of the Scale Problem?*

The a, b, c of the scale problem are:

a. *Atmospheric Ozone Depletion* (which is allowing harmful ultraviolet radiation to reach the earth's surface).

b. *Biodiversity Loss* (which is removing species 100 to 1000 times faster than at any previous time in the history of the planet).

c. *Climate Change* (which is threatening to disrupt global climate stability upon which all life depends).

Any one of these major global system changes would be a challenge in itself, but they are occurring together, and each is affecting the other. There are many other life support systems which are threatened in addition to these better known ones.

*What is Causing Scale Problems?*

Scale problems are caused by the physical size of the global economy; the sheer volume and toxicity of the mass of physical materials that are moved or transformed in normal economic production cycles are disrupting the ecosystems that provide global life support services. The physical size of the global economy continues to expand as economic growth is encouraged and facilitated by trade and other government policies at all levels.

*Are there Solutions to the Scale Problem?*

A variety of approaches to solving scale problems exist. Many have been implemented or are being tested in various countries and communities around the world. Because of the complex nature of scale issues, solutions are being explored via innovative economic and business ideas, creative public policy initiatives, sustainable technologies, and inspiring human values.

*What are the Obstacles to Implementing Solutions?*

Scale problems have slowly emerged as unintended consequences from ways of doing things meant to benefit humanity. Early economic theories were developed to improve the age old process of exchanging goods and services. Technologies were invented to make life easier. Material consumption has made living more comfortable and enjoyable for many. Profit making was encouraged to stimulate development, as well as to acquire riches. Belief systems evolved to help guide actions at particular points in history. The *"Histories of the rise and fall of civilizations"*[2] point out that the causes of the declines and disappearances of prior societies lay in those society's refusals to give up old habits and ideas that worked in the past. Big ideas get things done, but they change slowly, sometimes too slowly for societies to adapt to changing realities. Political histories also point out

[2] Taintner, 1988; Diamond, 2005

11

the roles of small elites who have vested interests in the status quo, and the influence they bring to bear in maintaining the old habits. It should be noted that there is no consensus about sustainable scale, which is a major reason to develop and share a wide awareness.

*How can these Obstacles be Overcome?*

Reviewing the existing and proposed solutions for dealing with scale problems reveals a wide array of opportunities for:

- an ecologically sustainable world.
- which would be considerably more just.
- with stronger and more vibrant communities.
- with considerably more time for leisure and individual pursuits.
- with less distress and social problems.
- and with significant opportunities for creative challenges to be undertaken.

*Who is interested in the sustainable scale?*

Mother nature is continually affected by problems which are inadequately addressed by decision makers whether they are representatives of government, deep states, businessmen, civil society influencers. The aim is to keep interest alive and continue to make generations aware, to disseminate information and follow day by day which options and solutions are within everyone's reach.

# CHAPTER I
## Understanding Scale

We live in a booming social and economic reality that in two centuries has seen an unprecedented increase in the rate of urbanization. With this trend in 2050 the population residing in the cities will have increased by 75%; we talk about over two billion people to host. How many natural resources we will need to make available to fill this gap? How much energy, materials, infrastructures, structures, social supports we will need to be efficient?

Geoffrey West, a theoretical physicist of British origin and an internationally renowned academic, says that *"the future of the planet is the future of cities and vice versa"*, so to deepen the knowledge of the developments of urban agglomerations from a qualitative and, above all, quantitative point of view, it would allow us to draw an almost real horizon on the future of the cities and the population that lives there.

He also refers to the second principle of thermodynamics when he says that *"the consumption of energy for the development of cities (creation of order) involves a growth in disorder and entropy"*. Studying cities is therefore a scientific subject.

All think considered, human civilization is currently experiencing a period of entropy, it tries to resolve by continually extracting energy, resources, creating innovation, maintaining and repairing where it succeeds and where it is capable of. This situation and the reactions of human civilization are the starting point for discussing everyone's life: aging, mortality, resilience and sustainability.

This applies to the single organism or single species as much as to the social or economic aggregates.

It is now clear that most of the problems that humanity faces are an expression of pollution and global warming generated above all by the urbanization that continuously feeds them. In this sense, cities can also be the solution, because they have energy, ideas and creative force in them: man.

Here, in the cities, the different economic systems established in the different places of the planet and the different interests (global warming, energy, public health, ozone hole, climate change etc.) actually represent a sort of entropy in continuous evolution that interacts and collides heavily with natural rules.

All this to say that there is a way to examine and perhaps resolve the issue. We must therefore appeal to *biology*, the only tool in man's hands that can allow him to answer the many questions regarding these complex models.

To examine all the phenomena with a magnifying glass, to be able to interpret them, to see how things change on the basis of their size and exploitation, we must then use a *scale* that will allow us to compare the values.

*"Viewed through this lens* - says West - *cities, businesses, plants, animals, our body and even tumors display a striking similarity in the way they are organized and function."*

On closer inspection, cities are built in the image and likeness of man.

Road networks, electric cables, sewer pipes, water and telephone networks are all united by the same physical and mathematical principles. These principles regulate flows in universal networks quite similar to those found in our brain networks or in the blood vessels of our body. There is in fact a certain modularity and, therefore, a consequent scalability. In fact, all models and all economic variables are scalable: no discovery so far.

What is interesting, however, are the formulas that regulate the growth of living beings - of all species - also applicable to cities or to larger or smaller interest groups, since the underlying mathematical and physical relationships have a characteristic of universality.

We will talk about it in detail in the following pages.

*Brief Overview:* Sustainable scale is about how the physical size of the global economy has grown so much that it no longer allows critical ecosystems to provide the level of life support services we depend on.

*Basic Concepts:* A few basic concepts provide an important perspective on sustainable scale.

*Scale Levels:* Sustainable scale issues can arise at local, regional or global levels. We focus on global scale issues: all are connected and important.

*Scale Categories:* Scale issues are all about sustainability - whether economic activities allow ecosystems to provide the level of services we depend on. Certain scale benchmarks are important for scale relevant policies.

*Measuring Scale:* A variety of ways of measuring sustainable scale are being developed. Although they use different methodologies, they all indicate we are in trouble.

*Moral Approaches:* Sustainable scale is related to the spiritual concept of human stewardship of nature.

*A Scale Synthesis:* Sustainable scale is about whether the human enterprise will thrive or collapse.

# A Brief Overview of Sustainable Scale

*The Basic Concept*

The concept of scale is about appropriate size relative to some standard. Sustainable ecological scale is about the appropriate size of the global economy in relation to the functioning of global ecosystems, which both contain and sustains the economy.[3] All economic activity degrades ecosystems, interfering with natural biological and biogeochemical processes that are critical to various life support services. The level of economic activity 250 years ago was small enough that the interference with ecosystems was sustainable. The current scale of human economic activity is unprecedented, interfering with global ecosystem functioning, is ecologically unsustainable, and may have potentially disastrous consequences for our civilization.[4]

*Contrast between Then ...*

In 1750 the physical size of the global economy was relatively small compared to the capacity of ecosystems to both provide resources and to absorb wastes. At that time the human population was less than one billion people, economic growth was present but not a dominant global policy priority, resources used by humans were primarily renewable, and wastes were largely biodegradable, non-toxic and dispersed locally. Ecosystem degradation which did occur was localized, and was usually restored through natural rege-

---

[3] Daly, Herman. Beyond Growth: the Economics of Sustainable Development, Boston: Beacon Press, 1996 Daly, Herman and J. Cobb. For the Common Good: Redirecting the Economy toward Community, the Environment and a Sustainable Future. Boston: The Beacon Press, 1989.
Daly, Herman and J. Farley. Ecological Economics: Principles and Applications, Washington,D.C., Island Press, 2004.

[4] McNeill, J. R. Something New Under the Sun. New York: W. W. Norton Co., 2000.

neration when the human groups responsible moved on to other sites.

*... and Now*

There are now over 7 billion people on the planet, the globally dominant policy of economic growth is driving ever greater quantities of both renewable and nonrenewable resources through the human economy, and many of the wastes generated are both toxic and spread broadly across the globe. Exploitation of some resources has been so high that they are nearly, or in some cases totally, depleted. Ironically, many of these resources (such as certain fisheries and plants) are ones which should be renewable. Disruption of ecosystems is no longer restricted to local areas, although the number of local areas adversely affected is larger than ever. The ecosystem degradation on many of these sites is so severe as to be unrecoverable over a human lifetime. Many of these sites are interspersed among human habitation. Site remediation, if undertaken at all, occurs at great expense.

*Unprecedented global threats*

Of even greater significance is the impact of the physical size of the global economy on the capacity of critical global ecosystems to meet human needs. The physical size and quality of the world's economic activity is now threatening the capacity of global ecosystems to provide sufficient basic life-support services for human well-being. These anthropogenic threats are unprecedented in the history of civilization and are occurring with an alarming speed. The A, B, Cs of sustainable scale − atmospheric ozone depletion, biodiversity loss, and climate change − are but the best known of these examples. Unfortunately, there are many more (we talk about in the Volume II).

The fact that human economic activities are affecting such large, powerful and robust global systems at all is cause enough for concern. Ecosystems change constantly due to natural phenomena and they will continue to function despite human interference. But human-induced changes of these global ecosystems could seriously degrade their capacity to provide life supports which our civilization depends on. At the very least, adapting to these human-induced ecosystem changes will be incredibly expensive. Attention to these issues is of crucial importance.

*Issue Largely Ignored*

Political decision makers are not adequately dealing with the potential seriousness of many of these challenges[5]. Almost all political parties around the world accept and support larger economies as an overriding policy priority. It is not surprising that the connection between the physical size of the economy and dangerous ecosystem degradation is avoided or resisted, especially when one considers the institutional, ideological and personal investments in this dominant policy. Solutions are likely to require different ways of thinking and new institutions regarding the relationship between economic activities and the level of ecosystem services human civilizations require.

*An Alternative Approach*

One such approach is emerging in the field of Ecological Economics.[6] The scale concept is a central issue in this new approach to macroeconomic theory and practice. Understand the scale it

---

[5] Speth, James. Red Sky at Morning. New Haven: The Yale Univesity Press, 2004.
  Speth, James (ed.). World's Apart: Globalization and the Environment. Washington: Island Press, 2003.

[6] Costanza, Robert (ed.). Ecological Economics: The Science and Management of Sustainability. New York: Columbia University Press, 1991.

means to understand these global challenges, and the exploration of effective and attractive solutions (we talk about in the Volume III).

*Connecting Vital Concepts*

The scale concept connects three of the most important issues of our global civilization in the 21$^{st}$ century – economic growth, ecosystem sustainability and social justice. Economic growth is easily one of the most defining characteristic of modern civilization. But along with the many wonders and comforts it provides, economic growth may also contain the seeds of our civilization's ecological and social decline. The sustainable scale concept directly addresses the relationship among these important and powerful issues, and provides a framework for ensuring that economic development allows ecosystems to continue providing the necessary support for a just and thriving human civilization.

# Basic concepts relevant to sustainable scale

Let's take a step forward highlighting the concepts underlie the idea of a sustainable scale.

*Material Throughput:* All economic activities involve the throughput of materials and energy that degrade the natural environment.

*Income:* involves living off what is renewed, rather than dipping into capital.

*Ecosystem Functions and Services:* Ecosystems are the basis of all human wealth, financial and otherwise.

*Natural Capital and Income:* Living off the natural income of ecosystems, without dipping into natural capital, is what ecological sustainability is all about.

*Thermodynamics:* The basic laws of physics indicate there are biophysical limits to how much throughput in the economy can be tolerated by the ecosystems we depend on.

## 1. Material Throughput

Thinking about sustainable ecological scale involves thinking about the amount of physical material moving through the global economy. This notion of *Material Throughput* helps us to connect the physical size of the global economy and the ecosystems it affects.[7] Each time we purchase a good or service we set in motion a chain of activities that has an impact on the physical world. Whether it is extracting resources from the earth, manipulating those resources in a production process, using the goods produced, or the eventual disposal of those goods as waste, physical material is being used and used up, energy is being expended and dissipated, and ecosystems are being degraded. There is

---

7 Daly, Herman and J. Cobb. For the Common Good: Redirecting the Economy toward Community, the Environment and a Sustainable Future. Boston: The Beacon Press, 1989.

little recognition that economic activity is impossible without some physical impact. Sustainable scale raises the question as to how much material throughput is possible while sustaining the ecosystem services that make economic and other important human activities possible.

The primary measure of economic activity, *Gross World Product* (GWP), is also a general measure of material throughput. The higher the GWP, the more material and energy use. This is true whether GWP is measuring steel production, information technologies or currency speculation. All economic activities require material throughput to a greater or lesser degree. An expanding global population, and growing levels of consumption, mean that the total amount of material throughput continues to increase, despite increased efficiencies in some areas. Economic activity and material throughput are inextricably linked, and one may serve as a rough proxy for the other. It is the scale of material and energy throughput in the global economy (along with its toxic qualities) that constitutes the main challenges for ecosystem capacities to serve human needs. Too much material throughput eventually exceeds the sustainable scale that allows ecosystems to continue providing the level of services human civilization depends on

## 2. Income

Another way of thinking about sustainable scale is in terms of income, which can be defined as the amount of something a community can use up without reducing the amount of capital from which it is derived, so that the same amount can be generated in future years.[8] By this definition income is sustainable - the same amount can be generated year after year. If more income is used up in a year than the capital available can generate, then

---

[8] Hicks, J. Value and Capital, 2nd edition. Oxford: Claredon, 1948.

the amount of capital will decline. Use at this level is no longer income, but depletion, and is not sustainable. It will only be a matter of time before the capital itself declines to zero, at which point there will be no more use. This is true not only for financial capital but also for *Natural Capital and Income*. When capital can no longer generate the same level of income as it has in the past, the sustainable scale of income used has been exceeded.

## 3. Ecosystem Functions and Services

*What are Ecosystems?*

Ecosystems are dynamic interrelated collections of living and non-living components organized in self-regulating units. Some degree of biodiversity exists in all ecosystems. An ecosystem is a unit because it has boundaries and can be distinguished from its surroundings. The living and non-living components affect each other in complex exchanges of energy, nutrients and wastes. It is these dynamic exchanges, both fast and slow, which provide eco-systems with their distinct identities. Because of these distinct features ecosystems themselves represent part of the earth's bio-diversity. The characteristic exchanges within an ecosystem are called ecosystem functions and in addition to energy and nu-trient exchanges, involve decomposition and production of bio-mass. The complex interdependencies which develop within or among ecosystems often create emergent properties, or characte-ristics that cannot be predicted from the component parts alone.

*Ecosystems Provide Stability*

Ecosystems are often characterized by one or more equilibrium states. An equilibrium state is a mildly fluctuating, relatively sta-ble set of conditions that maintain a population or nutrient ex-change at specific levels. Each equilibrium state is dynamic and undergoes periodic decline and resurgence depending on such factors as energy and nutrient inputs, predator-prey relationships (including diseases), or irregular disturbances. Even in its relatively stable condition an equilibrium state is dynamic in terms of the exchanges of nutrients and energy which occur, as well as the activities of its living components. As ecosystems incorporate more components into their functioning, for example as a result of energy, nutrient or waste accumulations, or the introduction of new species, their biodiversity can increase. More biodiversity increases ecosystem complexity, allows for the provision of more ecosystem functions, and may contribute to the occurrence of emergent properties.

*Ecosystem Functions Regulate Change and Stability*

Multiple stable states characterize most ecosystems. If disturbances or perturbations occur from either internal or external sources which tend to drive an ecosystem away from its current equilibrium state, then the ecosystem's regulatory feedback mechanisms work to maintain the current state, or to bring the ecosystem to one of its other typical equilibrium states. Which state is prevalent at any particular time has an impact on related ecosystems. Depending on which equilibrium state is prevalent, there will be more or fewer plants or animals in that ecosystem (or more of one type and less of another), more or less food available, more or less waste absorption, more or less nutrient cycling, or more or less energy.

*Ecosystems at Different Scales*

Ecosystems can be small or large. Our entire planet is covered with a variety of different, sometimes overlapping, and often interdependent ecosystems. Major global ecosystems are referred to as biomes. A tropical forest is a mid-sized ecosystem, which itself contains a diverse array of smaller ecosystems, and which in turn connects with global ecosystems. These layers of ecosystems are in dynamic interactions with each other, and influence which equilibrium state each maintains. Ecosystems are said to be "self-regulating" or "self organizing" because each contains feedback mechanisms which function to maintain the components of the system in one or other of its equilibrium states. An equilibrium state demonstrates the stability of ecosystems. Even in these stable states the components of ecosystems are in dynamic exchanges, and these exchanges involve the predictable build up of energy or materials which cycle the ecosystem either within a single equilibrium or between various equilibrium states.

 Ecosystems tend to cycle between these states of change and stability. Ecosystems of different sizes are interconnected and affect each other.  As ecosystems at one level ebb and flow between different stable states, they each create fast and slow cycles relative to their neighbors. Smaller ecosystems are generally characterized by faster cycles of change and stability, and larger ecosystems by slower cycles, with timeframes as long as a millennium or more.

*Disturbances*

In addition to the relatively predictable ebbs and flows of ecosystem cycles, less frequent and predictable external disturbances also occur (lightning induced fires sweep through a forest or grassland; a volcanic eruption spews tons of material into the atmosphere; a desert riverbed is flooded). These disturbances stimulate ecosystems to change within their existing equilibrium state, or if the disturbances are great enough the system may move to one of the other typical equilibrium states. When disturbances occur with regularity (although their timing and extent may be unpredictable, such as with fires) they are incorporated into the ecosystem's self regulating mechanisms. These adaptive mechanisms may either provide protection against the disturbance (eg. fire resistant bark) or rely on the disturbance to maintain itself (eg. fire induced bursting of pods to release seeds).

*Resilience Supports Stability*

An ecosystem is described as having *resilience* to the extent that it is able to return to its current equilibrium state following a disturbance. Ecosystems that have evolved over long periods of time have incorporated numerous adaptive mechanisms to various disturbances which provide them with the resilience to maintain their structures and functions, and to cycle between typical equilibrium states. Ecosystems which have survived for long periods of time have done so because they have been successful at maintaining their own structures and functions, as well as adapting to the fast and slow cycles of interconnected ecosystems, and to the unpredictable disturbances that are an inherent part of nature. If a disturbance pushes an ecosystem beyond its resilience capacity to adapt, the system will change into a chaotic state; eventually, a new equilibrium state will emerge from its components. Because of the complexities involved, and our relative ignorance of how ecosystems work, it is impossible to predict the characteristics of such new equilibrium states.

*Ecosystem Dynamics Provide Life Support Functions*

It is the adaptive capacities of ecosystems that have provided both the stability and the enormous range of diversity on earth. It was the ecosystem involving the oxygen producing micro organisms which emerged billions of years ago that provided earth with its unique atmosphere. Continued evolution of these ecosystems allowed the protective atmospheric ozone layer to develop, making the planet's surface safe from UV radiation for more complex life forms to emerge. The evolution of forest ecosystems and ocean plankton contributed significantly to the development of the greenhouse effect which provides us with climate stability. Soil fertility is dependent on complex ecosystems of insects and micro organisms creating rich top soils by cycling nutrients from both decaying organic matter and deeper mineral-rich sources. The wonderful varieties of plants and animals we enjoy as foods all evolved from unique ecosystem environments with specific requirements of moisture, temperature and nutrients.

*When Ecosystem Functions Become Services*

The specific ecosystem functions that are apparently beneficial to human civilization are called ecosystem services. However, given the early stages of human knowledge regarding ecosystems, it would be both premature and imprudent to exclude any ecosystem functions from this category. Ecosystem services clearly provide life support services for both humans and other species. Our dependence on ecosystem services is complete but poorly understood. Even in simplistic economic terms, the value of ecosystem services is larger than the global economy[9]. Ecosystem services go beyond the direct economic benefits derived from exploitation of very specific ecosystem functions such as timber from forests. It is ecosystems' ongoing capacities to provide a stream of life supporting and life enhancing services that are vital to human well being. Many of these services are non-market services by virtue of their inherent characteristics (eg. both the atmospheric ozone layer, and the climate stability provided by the global carbon cycle, cannot be owned by anyone who can con-

---

[9] Costanza, Robert, C. Perrings and C. Cleveland (eds.). The Development of Ecological Economics. Brookfield: Elgar, 1997.

trol their use by others; both ownership and control are conditions for a good or service to be traded in a market.[10]

*Ecosystems Connected to Basic Beliefs and Values*

There are also many ecosystem services that are thought to have intrinsic value, for moral, ethical or aesthetic reasons. It is becoming increasingly important to understand the many different types of benefits that ecosystems provide for human well being, and to reconcile the various approaches to valuing these benefits.[11]

*Market Economy Threatens Ecosystem Functions*

Over the 20th century human activities have placed increasing demands on certain ecosystem services, particularly those affected by the market economy. For example, the earth's forest cover has been significantly reduced to provide wood; ores have been mined for both fuel and to build things; plants and animals have been domesticated, bred and commoditized to provide food. The success of the human species has affected virtually every major ecosystem on the planet. Humans are omnivorous, capable of establishing settlements anywhere on the planet's surface, and make extensive use of the earth's renewable and non-renewable resources. While ecosystem science may be relatively new, human domination of the earth's ecosystems has been gradually expanding ever since the beginning of agriculture; it has grown exponentially with the ready availability of fossil fuels over the last 150 years. For a civilization that has sought to dominate nature, our impacts on ecosystems all over the planet are clearly

---

[10] Daly, H. E. and Farley, J. Ecological Economics: Prinicples and Applications, Island Press, Washington, D.C., 2004.

[11] Ackerman, F. and Heinzerling, L. Priceless: On Knowing the Price of Everything and the Value of Nothing, The New Press, New York, 2004.

indicating that we are very much a critical feature of these ecosystems which we are just beginning to understand.

## 4. Natural Capital and Income

The concept of *natural capital* refers to the source or supply of resources and services that are derived from nature. Forests, mineral deposits, fisheries and fertile soil are some examples of natural capital. Air and water purification are just two of many services.

*Natural Income* is the annual yield from such sources of natural capital - timber, ores, fish and plants, respectively, relative to the examples above. The point at which the amount of natural income used up reduces the capacity of natural capital to continue providing the same amount of natural income in the future, is the point at which sustainable scale has been exceeded.

*Natural Income: More than Resources*

Natural resources are not the only type of natural income which flow from ecosystems. A variety of *ecosystem functions* are also provided. Forests, for example, are not simply wood production units. They also prevent soil erosion, absorb rain water and provide flood control, they provide habitat for a diversity of plant and animal species which may serve as foods or medicines for other species, they absorb the natural wastes of these diverse life forms, they generate oxygen and sequester carbon from the atmosphere, they affect the microclimate of their area, they are a key component of the hydrologic cycle, as well as providing aesthetic enjoyment and spiritual inspiration. These forest ecosystem functions evolved to maintain the overall health of the forest environment and the creatures in it. Ecosystem functions are another form of natural income derived from the same natural capital of the forest ecosystem that generates timber for economic use.

Ecosystem functions that have particular value to humans are called *ecosystem services*.

*Services from Natural Capital*

There are four general services provided by natural capital, each of which need to be considered from the perspective of criticality:

- *Provisioning Services* - which provide resources used in production (timber, fish, etc.).
- *Regulating Services* - which regulate ecosystem processes, such as decomposition of organic wastes, cleansing of the air (by oxidation, etc.).
- *Cultural Services* - providing benefits of a spiritual, aesthetic, recreational or psychological nature; giving meaning to place, etc.
- *Supporting Services* - which regulate processes necessary for all the other ecosystem services.

*Resource Extraction Destroys Ecosystem Services*

This distinction between the resource (*stock*) and service (*fund*) functions of the same ecosystem is a key concept in understanding the relationship between the physical size of the economy and the ecosystem's capacity to support life. The same ecosystem provides both resources and services as forms of natural income. But only resource extraction, which generates large amounts of material throughputs, is identified in economic theory and practice. Lumber has a market price; erosion control or microclimate impact from the same forest likely do not. Consequently, most policy focus is on lumber production rather than on the full range of natural income provided by ecosystem services. It is the magnitude and pervasiveness of these ecosystem service losses, rather than resource depletion, which are the greatest threats to life on the planet.

*Natural Capital can disappear as Financial Capital thrives*

The natural income, or broad range of services, that flow from the same ecosystem all have different sensitivities to external disruption. When a given amount of timber is harvested from a forest, the ecosystem services of water retention and biodiversity habitat are differentially affected. It is possible that one or more of the many ecosystem services of a forest have passed a critical threshold of irrevocable collapse, even after timber is still harvested from the same forest. When sustainable scale is exceeded in such a manner, market prices of timber are clearly inadequate indices for signaling the loss of ecosystem services. Both the natural capital and natural income of specific ecosystem services can be depleted while financial capital continues to flow - even if only for a limited time.

*Destroying Ecosystem Services Destroys Sustainability*

Sustainable scale is exceeded for an ecosystem service when the rate of resource depletion reduces the capacity of natural capital to provide in the future the natural income it yielded in the past. Whether the process is with money or natural capital, as long as the draw down exceeds the rate of replenishment, the amount available will eventually shrink to zero - sustainability is destroyed. Thinking about sustainable scale forces us to focus at least as much on ecosystem services, and the natural income they provide, as on resources. Because natural capital is excluded from economic theory and practice, these vital, life supporting sources of natural income essential for sustainability, are considered to have no market value and are therefore ignored.

*The Competition for Natural Capital*

The exchanges of energy and nutrients among the planet's biotic and abiotic components created the earth's unique capacities to support complex life forms. Our planet's web of life depends on solar energy and exchanges of natural capital in life sustaining processes. As the human species evolved, spread around the globe, and began increasing the level of material throughput via technology, it began competing with other species and natural processes for the available, and finite, amounts of natural capital available. The success of this human activity is leading us to out-compete the natural world upon which we depend. Thinking about sustainable scale helps us think about the appropriate balance between the human use of natural capital, and the rest of nature's use of the limited amount available.

*Which Natural Capital Is Critical?*

The use of natural capital for the production of human artifacts is essential for the goods and services we need for our survival and well being. Therefore the destruction of some natural capital is an integral part of economic activity. A vital question is "how much natural capital must be preserved so that the level of natural income of both goods and services does not decline?" Another way of framing this question is in terms of "what aspects and amounts of natural capital are critical to human well being?" This is another way of approaching the sustainable scale issue.

*What makes a Function or Service Critical?*

There are three key characteristics that make an ecosystem function or service critical:

- *Non-Substitutability:* There is no substitute for it, either natural or man made (protection from solar radiation; climate stability, etc.).

- *Irrevocable Loss:* Its loss would be irrevocable, that is, if degraded beyond a certain level it would not recover in a meaningful human timeframe (biodiversity loss, toxic wastes, etc).
- *High Risk:* Its loss would constitute a considerable risk to human well being (eg. climate instability unprecedented in the history of human civilization, etc.).

*Identifying Critical Natural Capital*

Several major research projects are underway to identify what specific aspects of natural capital are critical, at both the global and regional levels.

## 5. Thermodynamics

The laws of thermodynamics are some of the most fundamental and powerful of all the laws of physics. They help clarify the finite nature of the biophysical world in which our economy operates, as well as the way material throughput in our economic activity often degrades critical global ecosystems.

The *first law of thermodynamics* states that energy cannot be created or destroyed, but only changed in form. The *second law of thermodynamics* deals with a fundamental fact of the transformation process. It basically states that whenever energy is transformed, some energy is converted to heat and dissipated, and therefore no longer able to do work. This loss in the ability to do work, means that the quality of the energy is degraded.

A similar process occurs for matter; matter cannot be created or destroyed but it can be transformed. When materials are transformed, their unique and valuable structural properties are always somewhat degraded and are no longer available to be used in precisely the same way again.

For example, a piece of coal has a particular organizational structure in chemical/physical terms. Coal is valuable because of its *energy density*, the temperature at which it can be combusted and the amount of energy it gives off. When coal is transformed in the process of combustion, these unique and valuable characteristics are degraded, and the resulting ash and heat energy released to the environment can never again accomplish the same amount of work. Economic activity not only uses material in the process of production, it also "uses it up" so that it cannot be used again in the future. This law applies to all the material throughput upon which our economy depends. Reuse and recycling may extend the usefulness of some materials, but each use degrades their valuable characteristics.

By pointing out the finite nature of the material world and the degrading aspect of all energy and material transformation during economic activity, these laws of thermodynamics and entropy beg the question of how much degradation of resources can or should be tolerated as a result of economic activity. This is another way of asking what the sustainable scale of the economy can or should be in relation to the ecosystems which contain and sustain it.

# Sustainable Scale Levels

When we talk about sustainability we should keep in mind we are surrounded by thresholds and margins. Life in this planet is organized by *Mother Nature* within certain limits. Limits that unfortunately we have learned to overcome and not to consider.

*Introduction to Scale Levels:* scale problems can arise at many different levels, and problems at one level can spill over into other levels.

*Local Scale Levels:* are as old as humans, but today's local scale problems are unprecedented in their seriousness and pervasiveness.

*Regional Scale Levels:* are caused by the same dynamics as problems at the local level - use of non-renewable resources and overuse of renewable resources and sinks.

*Global Scale Level:* solutions for alleviating local and regional scale problems (trade or migration) are unavailable at the global level. Global scale problems require global institutions and solutions

## 1. Introduction to Scale Levels

*Scale Terminology*

The term *"scale"* has many different meanings. As used in the phrase "sustainable scale" it refers to the physical size of the economy relative to the ecosystems which contain and sustain the economy. As such it is a relational term, describing the relationship or proportionality between two entities – in this case, between the economy and ecosystems.

Another important use of "scale" has to do with spatial relationships and their extension in time or duration. This is how the term is used in the phrase "scale levels". Scale levels are about the spatial and temporal extension of economic and ecosystem activities. Considerations of sustainable scale require us to consider scale levels; unsustainable scale can occur at different scale levels.

*Choosing Scale Levels*

There are a variety of ways of categorizing scale levels – from the molecular to the global. The criterion might be a social construct such as a national or city boundary, or a biological construct such as a bioregion or a watershed[12]. Whatever criteria are used to categorize scale levels have a political impact with respect to what gets studied and whose priorities are reflected. Toxic leachate from a local dumpsite is a different level from greenhouse gas concentrations in the atmosphere; different parties caused or contributed to the problem, different parties are affected, and different parties are responsible for implementing solutions. Different scale levels also involve different temporal cycles, and the types and comprehensiveness of empirical information available.

*Scale Level Interactions*

Regardless of what criteria are used to categorize scale levels, a key issue involves the relationships between different levels; are the underlying principles and dynamics which characterize one level the same or different from those in other levels? How do activities in one level affect activities in other levels? Are sustainability criteria the same across levels? When our areas of interest are economics and ecosystem functioning there are clearly effects across levels that are important from the perspective of ecological sustainability (we will see in *Panarchy*)·

*Scale Level Categories*

The focus here is the global level. Global issues of sustainable scale are arguably the most challenging and serious issues human civilization faces. Consequently, little information will be found here regarding sub-global scale levels. However, this is not to suggest that

---

[12] The Conceptual Framework Working Group of the Millennium Ecosystem Assessment. Ecosystems and Human Well-being, Chapter 5: Dealing with Scale. Island Press, Washington, D.C., 2003.

sub-global scale issues are unimportant. We have unprecedented challenges to global ecosystems because economic activities at local levels have cumulative impacts at local and regional levels, as well as globally. Consequently, solutions to global issues of sustainable scale are needed at local and regional levels, but with a global perspective.

## 2. Local Scale Levels

The earliest environmental problems caused by people were local in nature. Many of today's environmental problems are also local, where "local" may refer to a municipal jurisdiction, a neighborhood or even a specific property, such as a dump or industrial site. Air, water and soil pollution are all examples of local environmental problems. Reduced habitat caused by urbanization or industrial activities, perhaps even leading to extinction of *endemic* species, are other examples, as are local deforestation, soil erosion or loss of soil fertility due to industrial agriculture.

Many local environmental problems may create a nuisance, but this does not mean they are a sustainable scale problem. A local environmental problem becomes a sustainable scale problem if:
a local nonrenewable resource is being used.
a local renewable resource or sink is being used faster than it can be renewed.

In either case, sustainable scale can be exceeded at the local level: a local mine's ore may become depleted; a forest may be clear cut; a river may receive more effluent than it can purify; or a ground site may receive more emissions or toxic materials than it can flush out or breakdown.

While many local scale problems can be serious and endure for long periods, ecosystem *resilience* can often provide remedies. Sinks for air, soil or water emissions can be renewed by cutting or eliminating the emissions in some cases. Nonrenewable resources may be imported from another jurisdiction, as can renewable resources. Trade with other regions can help a local area that has exceeded sustainable scale for some source or sink. Wood can be imported if the local forest is lost. Food can be imported if the local water supply or soil fertility no longer supports local agriculture. Waste, toxic or otherwise, can be exported. Even the industrial operations that produce the waste can be exported. Among the other benefits trade may bring, one of them is to restore resource and sink flows that may have been degraded by unsustainable local practices.

*Local Examples*

There are also cases where sustainable scale has been exceeded locally. The Chernobyl nuclear reactor site is such an example. There are many other local sites around the world where the volume and or toxicity of emissions has been so great that they either have to be physically removed (generally at great financial costs), or the contaminated site has to be quarantined and taken out of active use (also an expensive proposition), and people have to be moved (as with the Love Canal in New York State, or the Sydney Tar Ponds in Nova Scotia). There are many such contaminated sites in most industrial countries, and the projected cost of cleaning them up is in the billions of dollars.

Local environmental problems are as old as man. In preagricultural times, if a local site was hunted or fished out the clan could move to another site with more resources. There are, however, a number of differences between these earliest local environmental problems and those of today. In the past, the nomadic lifestyle meant moving on to another site was "normal". Today, moving entire communities and abandoning a built-up infrastructure is virtually unheard of. There are also fewer uninhabited desirable sites to move to. Relying on trade to move resources or biocapacity or even waste is more likely than moving people.

In the past, human population groups were relatively small and the damage they could do to ecosystems, while considerable, was still small compared to the degradation caused by hundreds of thousands if not millions of people in a single city today. In the past, the wastes created were largely organic and while they might be noxious they would eventually decompose without doing permanent damage. Today, not only has the volume of our wastes increased dramatically, but also its toxicity. Some highly toxic wastes such as plutonium will remain dangerous for thousands of years.

## 3. Regional Scale Levels

A "region" may be defined, for example, by a national border, a trans-border trade zone, or a biologically defined *biome*. Data are most often available on a national basis, unless there is an international body regulating the area, such as the Joint Commission, a Canadian-US body managing the environmental quality of the Great Lakes.

As with local environmental issues, regional issues may be troublesome but not necessarily be sustainable scale problems. The latter occur under the same conditions that local and global sustainable scale problems occur: either non-renewable resources are being used; and/or renewable resources or sinks are being used faster than they can be renewed. Emissions causing acid rain cross borders expanding local scale problems to other regions.

*Regional Examples*

The Chernobyl nuclear reactor explosion caused both local and regional sustainable scale problems by spreading nuclear contamination over a large geographic area determined by wind patterns. Another example of a regional scale problem is the demise of the Aral Sea, from which water was used to irrigate cotton crops for export. The extraction of the water exceeded its replenishment rate, leading to the shrinking of the area covered by the sea, destroying fisheries and altering the ecosystems connected to the sea.

As with local level sustainable scale problems, exceeding scale at a regional level may be dealt with by importing bio-productive capacity from other regions. A good example of this is the importation of wheat by nations such as China. It has been argued that this is equivalent to importing water. China does not have the water to grow enough of its own wheat; by importing wheat it is effectively making up for having exceeded the sustainable scale of water use within its own territories.

## 4. Global Scale Level

We put the focus on global sustainable scale challenges. The focus on global issues is intended to clarify the pervasiveness and unprecedented nature of the challenges we face and to identify the global limits within which local and regional solutions are required. This approach is based on the assumption that understanding these limits is essential to first determining the seriousness of our current situation, and to designing local and regional solutions that will have a reasonable hope of developing sustainable scale at a global level in the future. Unless local solutions collectively address the global limits they will have little chance of ensuring that economic activities remain within sustainable scale. There are many examples of regional scale problems flowing over into the global level:

- depletion of local or regional water supplies.
- greenhouse gas emissions leading to climate change.

emissions of ozone depleting compounds responsible for thinning of the atmospheric ozone layer.

local or regional destruction of habitat for endangered species, to name just a few.

The local and regional accumulation of these problems has given rise to the global challenges to sustainable scale. The economic activities which give rise to these problems are carried out by people in specific communities around the world. The sheer physical size of this cumulative and globalized economic enterprise, and the lack of attention to the ecosystem impacts of how economic activities are designed and carried out, are what has given rise to sustainable scale problems at all levels.

*Why a Global Focus?*

The decision to focus on global issues of sustainable scale is based on:

the unprecedented nature of these problems.

the seriousness of the challenges they present (ie. irrevocable loss of critical ecosystem services).

the need to identify the global parameters within which local and regional solutions are implemented.

to clarify that global problems require global policy solutions.

At the local and regional level, sustainable scale problems can generally be dealt with through trade - by importing resources and biocapacity from other regions, and exporting wastes and even dirty production. With global sustainable scale problems this option is not available; the global limits are the ultimate limits. The fact that we are currently exceeding some of these limits is cause for grave concern[13].

---

[13] Millennium Ecosystem Assessment. A Synthesis Report. Washington: Island Press, 2005. www.ma-web.org

The global level is also the level which receives the least attention. Local and regional economic and ecosystem problems are the primary focus of most business, government and environmental groups; this is understandable as these groups focus on problems that are most immediate and noticeable. Global problems are more remote, and may be so slow to emerge that they appear to be more easily ignored. However, these temporal effects can be illusory – by the time the problems emerge with sufficient force to demand attention, it may be too late to affect the slow cycles set in motion by local or regional activities many decades or even centuries earlier. In addition, attempts to solve local problems could make the global challenge even more difficult to resolve (eg. exporting toxic wastes). Global problems require global institutions and policies.

# Scale Categories

What are the categories of scale? How many categories has the sustainable scale?

*Sustainable Or Unsustainable:* Human economic activities are either sustainable or not. This dichotomy is more important than how unpleasant environmental problems appear.

*Sustainable Scale:* There are various levels of sustainable scale. Two important benchmarks are *Maximum Sustainable Scale* and *Optimal Scale*.

*Unsustainable Scale:* There are various levels of unsustainable scale; resilience may allow recovery from low levels of unsustainable scale. If the maximum scale is exceeded there is no recovery of the ecosystem functions destroyed.

*Scale Policy Implications:* Sustainable is better than unsustainable; current unsustainable practices should be fixed immediately; maximum scale is to be avoided; optimal scale is a policy priority.

## 1. Sustainable Scale: Throughput Less Then Regeneration

*One or the Other: Sustainable or Unsustainable*

The amount of throughput in the global economy is either sustainable or unsustainable. All economic throughputs degrade ecosystems to some extent; the key issue is whether the rate or total amount of throughput is greater than the natural regeneration of affected ecosystem functions. If the impact of economic throughput degrades ecosystems faster than they can regenerate, there is an inevitable loss of ecosystem functions; this is unsustainable. If the impact allows the ecosystems to totally replace the resource or sink functions it interferes with, then there is no loss of function and they can endure; this is a sustainable scale. Economic throughput either degrades ecosystems faster or slower than they can regenerate; it draws down either more or less natural capital than ecosystems reproduce. This is a basic feature of ecological sustainability. Whether or not economic activities are sustainable (in terms of the ecosystems that contain and sustain them) has only become a relevant policy issue as evidence accumulates that current levels of economic throughput have crossed the sustainability threshold, and are continuing to increase.

## 2. Sustainable Scale

### Levels of Sustainable Scale

There are many levels of sustainable throughput. Available evidence indicates that up until at least the industrial revolution the level of throughput in the global economy was sustainable. Certainly, the level of hunter-gathers' throughput was sustainable, and considerably lower than the level of the throughput in the 16th century, which was also sustainable. No global ecosystems were challenged by the level of economic throughput until the 19th or 20th century[14]. Some levels of throughput provide very little in terms of material goods and services and some provide considerably more. Our challenge is to provide enough material goods for human well being while remaining within sustainable scale. Two conceptual levels of sustainable throughput, or benchmarks, are of particular interest:

- *Maximum Sustainable Scale* is the highest level of material throughput that remains sustainable; that is, where the rate of throughput is theoretically identical to the rate of regeneration. Any further increase in throughput becomes unsustainable. Maximum sustainable scale is determined by the biophysical limits of the ecosystems affected by the throughput in question. This conceptual notion of maximum sustainable scale is difficult to quantify

---

[14] Mc Neill, J. R. An Environmental History of the Twentieth Century World: Something New Under the Sun, W.W. Norton and Company, New York, 2000.

precisely, but is nonetheless an important conceptual tool in understanding the dynamics of sustainable scale.

- *Optimal Scale* is a level of material throughput within the sustainable range ( i.e. where throughput is less than regeneration), which provides the most benefits relative to costs, where the notions of benefits and costs includes ethical and social as well as economic values. Also included in optimal scale is the notion of a safety margin in terms of maximum sustainable scale. Optimal scale is therefore determined by socio-political limits, in addition to biophysical limits of ecosystems.

*Dynamics of Sustainable Scale*

There are some generalizations that can be made about the various levels of sustainable scale:

- within the range of sustainable scale, the lower the level of economically driven material throughput, the less risk there is of inadvertently exceeding maximum sustainable scale ( i.e. of moving into the unsustainable range). The lower the throughput the higher the margin of safety regarding the maintenance of critical ecosystem functions.

- within the sustainable range, the higher the level of economically driven material throughput, the higher the risk of inadvertently exceeding maximum sustainable scale (i.e. of moving into the undesirable range of unsustainable scale). The higher the level of throughput, the lower the margin of ecosystem safety.

- within the sustainable range, the more material throughput, the more material goods and services there are available for human use and enjoyment.

- within the sustainable range, the relationship between levels of throughput and safety regarding ecosystem functions is likely non-linear; that is, discontinuities will occur and safety levels will be difficult to predict.

## 3. Unsustainable Scale

*Levels of Unsustainable Scale*

There are also many levels of unsustainable scale, ranging from a level where throughput is only slightly higher than regeneration, to a level where throughput exhausts or destroys the ecosystem's capacity to regenerate, where resilience is exhausted. Obviously, these different levels of unsustainable scale represent different levels of threats to the well being of humans and other species. Unsustainable levels of throughput that are only slightly higher than maximum sustainable scale are less dangerous than levels of throughput which greatly exceed the rate of regeneration. If the level of throughput is so much higher than the rate of regeneration that it overwhelms the ecosystem's capacity to continue functioning, the basic ecosystem dynamic is changed and a new equilibrium established, then *Maximum Scale* is exceeded. The concept of maximum scale represents the point of no return, the level of throughput where the ecosystem functions upon which we depend, are no longer available. Maximum scale is defined by the biophysical limits of affected ecosystems. Such a level of throughput may be difficult to identify empirically with precision; the concept nonetheless serves as a useful reminder of the inevitable consequences of too much throughput relative to what ecosystems can process.

*An Invisible Threshold*

The shift in the global economy from a sustainable to an unsustainable scale of material throughput was hidden from view and daily experience. The same amount of benefits from nature (eg. timber, fish, biodiversity, etc.) could be experienced even after this threshold was crossed. The difference was that once we passed maximum sustainable scale, the benefits were being derived from a drawdown of natural capital, rather than the annual renewals of natural income. Our lack of a natural capital accounting system meant we could not distinguish this silent shift. Because the capital fund was so large relative to the demands made on it, we could enjoy the same drawdown without realizing a threshold had been crossed.

*The Inexorable Slide*

However invisible, once sustainable scale was exceeded, our growing levels of material throughput (and their increasing toxicity) meant we were on an ever more rapid slide toward a catastrophic point of no return. Once sustainable scale is exceeded, it is just a matter of time before ecosystems are degraded to the point where they can no longer generate services to support human well being. The fact that our levels of material throughput continue to grow means we will reach this catastrophic state that much sooner. Available evidence indicates sustainable scale has been exceeded in several critical areas.

*Ecosystem Resilience Buys Time…*

Ecosystems have the remarkable capacity to regenerate themselves if left undisturbed. A forest that burns from a lightening strike will eventually grow back; over-fished stock will regenerate if sufficient numbers remain and their habitat is intact. This characteristic *resilience* of ecosystems means that the inexorable slide toward a catastrophic point of no return can be slowed and reversed, bringing the level of material throughput causing the degradation back to a sustainable level.

*But Not Forever…*

Ecosystem resilience also has its limits. Resilience operates over the entire range of throughput levels, but the more material throughput that must be dealt with, the more difficult it is for the mechanisms of resilience to regenerate natural capital. Resilience declines when sustainable scale is exceeded; the higher the level and duration of throughput within the unsustainable range, the weaker resilience becomes. When biophysical limits are reached irrevocable losses will occur and resilience will be extinguished.

*Dynamics of Unsustainable Scale*

The dynamics of unsustainable scale include the following:

- within the unsustainable range, the lower the level of economically driven material throughput the less the degradation to ecosystem functions (i.e. the easier it is for ecosystem resilience to restore function).

- Within the unsustainable range, the higher the level of material throughput the greater the degradation to ecosystem functions (the higher the level of throughput the less likely is resilience to restore functioning).

- within the unsustainable range, regardless of the level of material throughput, the longer throughput remains unsustainable, the greater the degradation to ecosystem functions (the longer unsustainable throughput endures the more resilience is weakened).

- within the unsustainable range, regardless of the level of material throughput, if the throughput endures long enough, the capacity of affected ecosystems to continue functioning will eventually be destroyed (i.e. resilience will be exhausted and maximum scale will be exceeded) and a new equilibrium will be established.

- within the unsustainable range, the relationship between the level of material throughput and the level of ecosystem degradation is non-linear; small increases in the amount, or duration, of unsu-

47

stainable throughput can result in high levels of degradation to ecosystem functioning.

## 4. Scale Policy Implications

*Policy Implications of Scale Dynamics*

Many policy implications flow from the dynamics associated with sustainable and unsustainable scale levels:

- *Sustainable is Better than Unsustainable:* It is more desirable for economically driven levels of throughput to be sustainable than unsustainable (as sustainable levels maintain ecosystem functions that economic and other vital life support activities depend on).

- *Fix It Fast:* If levels of economically driven throughput are unsustainable, efforts should be made to reduce throughput levels, and return to sustainable levels as soon as possible (as the longer we operate at unsustainable levels the more we degrade ecosystems and their ability to regenerate, and the more difficult and costly will be the process of restoring sustainable scale).

- *Avoid Maximum Scale:* Exceeding maximum scale represents an irrevocable loss of critical ecosystem functions that threaten human well being and survival.

- *Optimal Scale the Policy Priority:* If a sustainable level of throughput is close to maximum sustainable scale there is always the possibility that some unforeseen natural or anthropogenic events could push the level of throughput to an unsustainable level. It is desirable to manage our level of economic throughput so that such an inadvertent transition to unsustainable scale does not occur. This involves assessing risks, as well as other ethical issues, for current and future generations of humans and other species. Also to be considered is the level of material goods and services desirable for human well being. Broad participation is needed to determine the socio-political priorities of optimal scale.

*Optimal Scale: the Policy Priority*

The concept of optimal scale incorporates ecological sustainability as well as social and ethical concerns. It identifies the level of economically driven material throughput that is most desirable from these multiple perspectives, each one fundamental to human well being – the level of throughput which provides the greatest economic, environmental and social benefits. One of the social and ethical considerations involves the responsibilities of the present generation to future generations of humans and other species. One aspect of such considerations is the safety margin that is acceptable in terms of inadvertently exceeding sustainable scale, and thereby slipping into unsustainable practices.

*Creating a Safety Margin: Accepting Limits*

Unsustainable levels of throughput are dangerous because of the degradations they impose on critical life support systems; they are to be avoided. But the biophysical limits which determine ecological sustainability are complex, dynamic, and non-linear, making it difficult to describe the boundary between sustainable and unsustainable throughput with great precision. This means limiting the level of material throughput below the level that is maximally sustainable, to create a safety margin. But the less material throughput we have, the fewer material goods and services we have to enjoy. The current economic paradigm assumes that more material goods and services mean greater human happiness and well being. Optimal scale recognizes that material goods are only one of the determinants of human happiness and well being.

*Managing Global Material Throughput*

The concept of sustainable scale provides a conceptual framework for managing the level of economically driven material throughput so that it remains within the safe, sustainable range. This framework assists us to get the most benefits from economic activities, as well as from the non-market benefits of life support ecosystems. As a relational concept (throughput relative to optimal regeneration) a scale perspective requires a new way of thinking about how to integrate economic, ecological and social/justice priorities and activities. What we have learned to do effectively (if not necessarily efficiently) in the last 160 years is how to drive throughput in economic activities. We have yet to learn how to mange throughput levels to remain within the biophysical limits of ecosystems, and the ethical limits of our highest common aspirations.

# Measuring Scale

*"Sustainable development is development that meets the needs of the present, without compromising the ability of future generations to meet their own needs."*[15]

Unfortunately development is driven by one particular need, without fully considering the wider or future impacts. Sustainable development is to set up sustainability (ecological sustainability, social sustainability, economic sustainability, institutional sustainability). When we talk about oil and gas or mining they are commonly relatively against ecological sustainability, such as biodiversity, habitat, etc. Then what could we conduct to optimize mutual benefit and sustain all those components of sustainability? Can we measure it or is it only a concept without a conclusion?

A variety of measures and concepts are evolving, each bringing a unique perspective to some aspect of monitoring sustainable scale:

*Carrying Capacity:* biological science describes the parameters that determine the rise and collapse of living populations.

*The IPAT Equation:* Impact * Population * Affluence * Technology; linking various human activities with ecological sustainability.

*Limits to Growth:* a variety of future scenarios project an unsustainable future (accurately to date) unless we change our ways.

*Ecological Footprint:* a standardized way to measure what has been immeasurable, contentious, and often misunderstood: ecological limits. It does this by comparing human resource demand to nature's ability to renew.

---

[15] Definition from *"Our Common Future"*, also known as the Brundtland Report.

*Critical Natural Capital:* some ecosystem services are more critical than others - they are necessary for our survival, cannot be replaced, and once lost are gone forever.

*Material Flow Analysis:* this methodology focuses on the sheer volume and types of material throughput that characterize various national economies; it highlights where the flows originate and where they end as wastes.

*Millennium Ecosystem Assessment:* this latest and most comprehensive scientific review of global ecosystems clearly documents that nature's services may not be there for us much longer.

*Panarchy:*[16] this comprehensive conceptual framework connects different scale levels, and describes dynamic adaptive cycles through which ecosystems naturally evolve. It has proven a useful framework for many practical applications at the local and regional levels, and is now being examined at the global level.

*The Natural Step:* focuses on an action plan for organizations to follow to achieve sustainability; its basic principles parallel those a sustainable scale perspective.

*Ecological Economics Perspectives:* Sustainable scale is about how economic throughput impacts the ecosystems which contain and sustain the economy. It can be viewed in terms of stocks or flows, and both approaches add to our understanding.

## 1. Carrying Capacity

*Basic Biology*

One of the earliest concepts related to the issue of scale is that of carrying capacity.[17] Biologists define carrying capacity as the

---

[16] Gunderson, Lance and C. S. Holding. Panarchy: Understanding Transformations in Human and Natural Systems. Washington: Island Press, 2002.

[17] Nebel, B.J., and R.T. Wright. Environmental Science: The Way the World Works. Seventh Edition. Prentice Hall, New Jersey, 2000.

maximum population of a given species that can survive indefinitely in a given environment. It was originally applied to relatively simple population-environments such as the number of sheep or cattle that could be maintained on grazing land without degrading the land so that it could no longer support the animals. It depends on the conditions and resources available in the specific area, and the consumption habits of the species considered. Because both what is available in the area, and the consumption habits of the species change over time, carrying capacity is always changing. Carrying capacity is a measure of sustainability within these changing conditions.

*Two Patterns*

Many animal species have been studied with respect to a specific area's carrying capacity. Starting from a low population level there are two quite different patterns which describe how various species reach carrying capacity, the sigmoid and peak phenomena. Populations which exhibit the sigmoid pattern increase rapidly while food and habitat are abundant, and then slow down as regulatory factors such as lower birth rate and reduced food availability come into play. As the rate of population growth slows down to zero, the population reaches a fairly stable level. This pattern is referred to as "*K*" (for constant) selected species. The other pattern of reaching carrying capacity is similar in the early stages when the population is still small. But in this situation the same regulatory factors do not come into play and the population increases rapidly to the point where it exhausts the resources upon which it depends. At this point, mortality becomes the primary regulatory factor, and the population collapses to a low level. When resources are replenished the population begins to rise again; this process is repeated in a boom and bust cycle. This is referred to as the "*r-selected*" species.

*Human Application*

The concept of carrying capacity was applied to human populations in the 1960's. It was noted that the consumption habits of humans are much more variable than those of other animal species, making it considerably more difficult to predict the carrying capacity of the earth for human beings. This realization gave rise to *The IPAT Equation* which pointed out that carrying capacity for humans was a function not only of population size, but also of differing levels of consumption, which in turn are affected by the technologies involved in production and consumption. There have been a large number of published estimates for the human carrying capacity of the earth; they range from a low of one half billion people to a staggering 800 billion. Many of these estimates are more ideologically based than determined by scientific principles.[18] These exercises demonstrate the complexity of developing useful estimates of the human carrying capacity of the planet, and the limitations of using the methodology which has been successful with non-human species. Various estimates of human carrying capacity also demonstrate in a general way the relationships among some of the major factors involved. Obviously, if consumptions levels per capita are higher, then a smaller population can be supported. If technologies increase or decrease overall consumption, then they also affect carrying capacity. Because the idea and methodology of carrying capacity were developed in the natural science of biology, they incorporate the notion of limits imposed by the earth's natural systems. Species can overshoot these limits (as with the *r-selected* species), and when they do, they collapse and risk extinction. The big question for human civilization raised by this application of carrying capacity to the human population is whether we will be a $K$ or $r$-

---

[18] Cohen, Joel. How Many People Can the Earth Support? New York: W. W. Norton, 1995.

*selected* species; whether we will reach a stable level that can be sustained for an indefinite period; or whether we will grow to a peak and collapse. Biological studies of various species provide us with some basic lessons to apply to the human condition, but new ideas and methodologies are needed to incorporate the added complexities of human technologies and culture. But carrying capacity tells us that the biophysical limits of our environment are key in determining how many human can survive at what levels of consumption.[19]

*Relation to Sustainable Scale*

The concept of carrying capacity is well rooted in biological science, and describes the rise and decline of plant and animal populations. It clarifies that there is a limit to the growth of any biological population, and identifies some of the parameters that determine the pattern of population rise and collapse. Additional layers of complexity occur for the human population in terms of the dynamics involved. Human choices are needed to ensure we imitate a *K* rather than an *r-selected* species.

*"...carrying capacity is determined jointly by human choices and natural constraints. Consequently, the question, how many people can the Earth support, does not have a single numerical answer, now or ever. Human choices about the Earth's human carrying capacity are constrained by facts of nature which we understand poorly. So any estimates of human carrying capacity are only conditional on future human choices and natural events."* Joel Cohen.

## 2. The IPAT Equation

*What is the IPAT Equation, or I = P x A x T?*

---

[19] Czech, B. Shoveling Fuel for a Runaway Train: Errant Economists, Shameful Spenders, and a Plan to Stop Them All. University of California Press, 2000, pgs 88-92.

One of the earliest attempts to describe the role of multiple factors in determining environmental degradation was the IPAT equation.[20] It describes the multiplicative contribution of population (P), affluence (A) and technology (T) to environmental impact (I). Environmental impact (I) may be expressed in terms of resource depletion or waste accumulation; population (P) refers to the size of the human population; affluence (A) refers to the level of consumption by that population; and technology (T) refers to the processes used to obtain resources and transform them into useful goods and wastes. The formula was originally used to emphasize the contribution of a growing global population on the environment, at a time when world population was roughly half of what it is now. It continues to be used with reference to population policy.

*New Insights*

In addition to highlighting the contribution of population to environmental problems, IPAT made two other significant contributions. It drew attention to the fact that environmental problems involved more than pollution, and that they were driven by multiple factors acting together to produce a compounding effect. Subsequent research using IPAT indicates that the assumption of a simple multiplicative relationship among the main factors generally does not hold – doubling population, for example, does not necessarily lead to a doubling of impact. Approaches which allow for different weighting to be assigned to each factor have been more successful in accounting for impact.[21]

---

[20] Commoner, Barry. "The Environmental Cost of Economic Growth." in Population, Resources and the Environment. Washington, DC: Government Printing Office Pp. 339-63, 1972.

[21] Chertow, M. R. "The IPAT Equation and Its Variants; Changing Views of Technology and Environmental Impact," *Journal of Industrial Ecology*, 4.4 (2001): 13-29. accessed at: http://mitpress.mit.edu/journals/pdf/jiec_4_4_13_0.pdf

Attempts to strengthen the predictive power of the equation have been made in terms of incorporating a variety of social, political and technical factors.[22] Some of these studies[23] enhance the equation's usefulness for policy development by indicating the variable contribution of different factors, who is responsible for various factors and therefore where resources might best be directed to reduce impact most effectively. However, making the formula more complex also makes it more difficult to apply.

*Limitations*

To date, IPAT applications have been limited to evaluation of a single variable measure of environmental impact, such as air pollution. For example, the Intergovernmental Panel on Climate Change has applied IPAT to studies of $CO_2$ levels.[24] The equation is helpful, to a limited extent, in assessing the contribution of different PAT factors to greenhouse gas (GHG) emissions. The report suggests that levels of GHG emissions for affluent countries increase with increases in affluence, while both population and level of affluence can be significant factors in GHG emission trends in poorer countries. Various applications have found that different types of impacts (eg whether $CO_2$ or $SO_2$ levels are being considered) relate differently to changes in population, affluence and technology.[25]

---

[22] Fischer-Kowalski, M. and C. Amann. "Beyond IPAT and Kuznets curves: Globalization as a vital factor in analysing the environmental impact of socio-economic metabolism." *Population Environment* 23.1 (2001): 7-47.

[23] Dietz, T. and E. A. Rosa. "Rethinking the Environmental Impacts of population, Affluence and Technology." *Human Ecology Review*, 1.1 (1994).

[24] IPCC. Special Report on Emissions Scenarios: a special report of Working Group III of the Intergovernmental Panel on Climate Change. Cambridge, UK: Cambridge University Press, 2001. http://www.grida.no/climate/ipcc/emission/050.htm

[25] Cole, M. A. and E. Neumayer. "Examining the impact of demographic factors on air pollution," *Population Environment* 26.1 (2004): 5-21.

*Relation to Sustainable Scale*

From a scale perspective, the IPAT equation does not help us to identify sustainable limits regarding either individual or composite environmental impacts. It does assist in our understanding of the general factors that increase or decrease environmental impact, but not the level of impact that exceeds sustainable scale. However, by highlighting the complex interplay among a variety of factors in creating an impact, the IPAT equation also demonstrates that there are multiple ways of reducing undesirable effects. It has been noted, for example, that different nations might focus on different factors to reduce their overall impact: more affluent countries could contribute most by reducing their level of consumption (A); many poorer countries could contribute most by reducing their population (P); other iper technological countries could make the greatest contribution by making their technologies more energy/recycling efficient (T). While there is some truth to this observation, it is also true that opportunities exist in most nations to make improvements in all three factors.

*In Summary*

The IPAT equation made a contribution to understanding the multiple causes of environmental impact, and it continues to be developed as a method for improving our understanding of these issues. It has not helped in identifying sustainable scale, but it is a useful framework to assist in thinking about ways of reducing environmental impacts by reducing various types of throughput.

## 3. Limits to Growth

*Projecting the Future*

In 1972 the book *"Limits to Growth"* was published and sent out shock waves around the world.[26] It contained a warning that by the year 2100 the world might be on a collision course with catastrophe if then current rates of growth in such areas as resource use, industrial output, food production and population expansion continued on their then current course. This warning was the result of computer modeling of a variety of future scenarios, based on different assumptions concerning the future state of the world applied to the best data available regarding various growth parameters. A variety of measures of human welfare were also included. The computer model tested the results of different assumptions about the future state of the world with respect to population, resource use, etc in terms of the impact on human welfare. The only computer scenarios which indicated human welfare could be sustained were ones in which growth was reduced.

*Warnings Ignored (And Denounced)*

The authors of the study were attempting to point out the consequences of continued growth in population, resource use, pollution, and so on, based on then current trends. Their intent was to stimulate debate and discussion about how to plan for the projected overshoot of global carrying capacity which their modeling revealed. The concept of overshoot is similar to that of exceeding sustainable scale. Considerable debate did occur and the work was criticized as unscientific, inaccurate and overly pessimistic. It was largely ignored by policy makers who continued pursuing economic growth as a primary goal. Whenever the limits to growth argument came

---

[26] Meadows, Donella, J. Randers and D. Meadows. Limits to Growth. New York: Universe Books, 1972.

up as time passed it was denounced as proven wrong by the facts of continued growth. Common criticisms of the limits to growth warnings were that technological innovation and market signals would allow growth to continue. It was argued that as resources became scarce their market prices would increase because of scarcity and this would reduce demand for them. Technological innovation would then find substitutes so that growth could continue.

*Persistent Confirmation of Impending Collapse*

The study team published both a 20 year[27] and a 30[28] year follow up, adding measures and making improvements in their computer simulation model. These analyses came to the same conclusions as the original study – that continued growth would lead to overshoot and catastrophe for human civilization. In their original study in 1972 they warned that overshoot was a possibility; in the 1992 report *"Beyond the Limits"* they argued that overshoot had already occurred in a variety of areas, and that their original warning were even more urgent. They pointed out that once overshoot of carrying capacity has occurred it will inevitably lead to collapse unless the process is reversed. They also pointed out that further delays in recognizing and dealing with the overshoot issue would actually reduce the options available for returning the world to a sustainable state.

In their 30 year update the study team point out some lessons they feel have been learned from their computer simulations, many of which explore assumptions about both technological innovation and resource substitution:

[27] Meadows, Donella, J. Randers and D. Meadows. Beyond the Limits. White River Junction, VT: Chelsea Green Publishing Co., 1992.

[28] Meadows, Donella, J. Randers and D. Meadows. Limits to Growth: The Thirty Year Update. White River Junction, VT: Chelsea Green Publishing Co., 2004.

- If one limit is removed but growth continues overall, then another limit will be encountered; they point out that there are layers of limits which are likely to unfold in successively multiple ways. Continued growth will only accelerate this process.
- If a society is in fact successful in putting off limits through economic or technical adaptations, it runs the risk of later exceeding several limits at the same time. What such a society runs out of is the ability to cope.
- Markets and technologies are tools that serve goals set by society; if the primary goal is growth these tools will be used in service of growth.
- Adjustments by markets or technology also have costs, and as limits are approached these costs increase dramatically, making the adjustments unaffordable.
- Markets and technologies operate through feedback loops with information distortion and delays; such delays facilitate overshoot.

*Reducing Material Throughput Necessary for Sustainable Scale*

Even when very optimistic assumptions about technical innovation are made, limits are reached and exceeded. It is only when reductions in material and energy consumption are combined with technological change that the computer scenarios produce a sustainable state for the world. Despite the warnings inherent in their work, the study team indicates that operating within the limits of the earth's carrying capacity is both possible and can be attractive. They also point out, however, that the longer growth continues to exceed these limits, the less attractive the options available.

*Relation to Sustainable Scale*

The Limits to Growth scenarios and the sustainable scale concept both acknowledge that the human enterprise has reached important limits and that we are in overshoot. Both approaches also suggest that operating the global economy on the edge of sustainability (i.e. at maximum sustainable scale) is dangerous and should be avoided. Both approaches examine the relationship between economic activity and ecosystem functioning; despite using somewhat different evidence and methodologies. Both approaches come to essentially the same conclusion: our economic activities may be destroying our civilization, and attention to this challenge is urgent. The Limits to Growth approach examines a variety of solutions to the issue of how our civilization might deal with the challenge of limits imposed by both non-renewable resources, and the biophysical limits of ecosystems. Computer simulation models are used to explore a variety of approaches to managing human affairs in such a way that reaching the limits could be avoided. The scenarios generated make it clear that the only solution is one which results in reduced levels of material throughput.

## 4. Ecological Footprint

*What Is the Ecological Footprint?*

The Ecological Footprint is rooted in the fact that all renewable resources come from the earth. It accounts for the flows of energy and matter to and from any defined economy and converts these into the corresponding land/water area required for nature to support these flows. The Ecological Footprint is defined as "the area of productive land and water ecosystems required to produce the resources that the population consumes and assimilate the wastes that the population produces, wherever on Earth the land and water is located."[29] It compares actual throughput of renewable resources

---

[29] Wackernagel, Mathis and W. Rees. Our Ecological Footprint. Gabriola Island, BC: New Society Publishers, 1996.

relative to what is annually renewed. Non-renewable resources are not assessed, as by definition their use is not sustainable.

The total "footprint" for a designated population's activities is measured in terms of 'global hectares'. A global hectare (acre) is one hectare (2.47 acres) of biologically productive space with an annual productivity equal to the world average. Currently, the biosphere has approximately 11.2 billion hectares of biologically productive space corresponding to roughly one quarter of the planet's surface. These biologically productive hectares include 2.3 billion hectares of ocean and inland water and 8.8 billion hectares of land. The land space is composed of 1.5 billion hectares of cropland, 3.5 billion hectares of grazing land, 3.6 billion hectares of forest land, and 0.2 billion hectares of built-up land. These surfaces represent the sum total of biologically productive hectares we rely on for our survival. They represent the earth's natural capital, and their annual yield represents our annual natural capital income.

*Ecological Overshoot Demonstrated*

Dividing the 11.2 billion hectares available by the global population indicates that there are on average 1.8 bio-productive hectares per person on the planet. The *2004 Living Planet Report* indicates that the actual usage was 13.5 billion global hectares or 2.2 hectares per person – more than a 20% overshoot.[30] The overshoot result indicates that our annual draw down of natural capital is liquidating natural capital income, as well as reducing natural capital itself. Such an overshoot is ecologically unsustainable. Time series of the global Ecological Footprint indicate that human activities have been in an overshoot position for approximately three decades, and the overshoot is increasing over time. Empirically demonstrating that

---

[30] Monfreda, C., M. Wackernagel and D. Deumling. "Establishing national natural capital accounts based on detailed Ecological Footprint and biological capacity assessments." *Land Use Policy* 21 (2004): 231-246.

ecological overshoot is now occurring by a significant margin is a major contribution to our understanding that we are exceeding sustainable ecological scale on a global level, and by roughly how much. The implications of these results are even more urgent when we realize that the Ecological Footprint is likely an underestimate of the actual demands we place on the earth's ecosystems.[31]

*The Footprint of Different Activities*

This measure can also be presented in terms of the types of products or services provided by the global hectares, for example, in terms of goods from crop lands, animal products, fish, forest products, built up areas, and energy and water use. Such analyses identify which areas are placing the greatest strains on ecosystems, and can help set policy priorities. Growth in animal products and energy use, especially of fossil fuels, are two areas that are rapidly increasing these strains.

*The Footprint of Nations*

Ecological Footprint looks at the total amount of global hectares that are required to support a particular population, regardless of whether those hectares are within the national borders where that population lives. It does this by considering the net consumption of the population (or activity) of interest, subtracting the global hectares used for export from those used for imports and production. The Footprints of individual nations vary considerably, from highs of near 10 hectares per capita for such countries as the United Arab Emirates, the United States and Kuwait, to lows less than 1 hectare per capita for such countries as Haiti, Somalia and Afghanistan. By comparing the Footprint measure with the actual bio-productive capacity of individual nations it is possible to determine if that country is in an ecological deficit (using more than it has) or has an

---

[31] Global Footprint Network. www.footprintnetwork.org

ecological reserve. The US, Japan, the UK and the United Arab Emirates are all in ecological deficit, using more global hectares than their own land mass provides. Countries with an ecological reserve include Australia, Mongolia, and Gabon.

Some, but not all, countries can run ecological deficits by appropriating bio-productive hectares from other countries. However, the global deficit represented by the 20% overshoot cannot be compensated for as there is only one planet available. These data highlight the intimate connection between ecological sustainability and just distribution, and the contribution of international trade to inequities in national Footprints.

*Sound Methodology*

The methodology for the Ecological Footprint is detailed but not overly complex. Data inputs are from publicly available national, international and private organizations. A variety of accounting assumptions are made, but they are explicit and always entail a conservative bias. Weaknesses in this pioneering endeavor have been acknowledged, many have been corrected, and others are being addressed with further research.

*A Policy Tool: The Footprint as an Indicator*

One of the many strengths of the Ecological Footprint is its immediate intuitive appeal. Along with its reasonable and continuously improving methodology, this appeal has led to its widespread use in a variety of settings, addressing national, regional, municipal and even individual footprints. The measure itself simply describes the size of the footprint for a particular population or activity. But its implication for policy and planning purposes has been recognized, leading to its use by several countries and municipalities to implement and monitor their sustainable development agendas. It has proven a useful research tool to explore the footprint of specific activities such as different modes of transportation or methods of farming. There is also an annual global footprint report that provides a useful overview across many specific areas.[32]

*Limitations*

The Ecological Footprint is not a precise measure of ecological sustainability. While it is perhaps the best estimate to date, it is important to recognize its limitations. In general, the Footprint underestimates the impact of human activities on the biosphere. Any applications of the Footprint methodology must keep this perspective in mind. Because it focuses on renewable resources, the Footprint provides limited information about most non-renewable resources and their impact on ecosystems (with the exception of fossil fuel impacts which it partially addresses). The concept of "global hectares" of world average bio-productivity is useful for looking at issues related to global Footprint. But individual applications refer to specific locations where there is an impact. These local areas may have bio-productivity rates different from the global average; where available, local data can be used. Another limitation is that the approach allows only general types of bio-productive areas to be identified (e.g. cropland, forests, etc). Specific ecosystems within these areas are not addressed. These limitations do not invalidate the Footprint, but do underline the importance of interpreting any specific application with these limitations in mind.

*Relation to Scale*

The Ecological Footprint is the closest empirical measure now available to estimate maximum sustainable scale. It captures the bio-productive capacity that is required to support a given level of material throughput, with current practices and systems of organi-

---

[32] "Living Planet Report 2004." World Wildlife Foundation. http://www.panda.org/downloads/general/lpr2004.pdf

zation. Maximum sustainable scale relates the physical amount of material throughput in economic activities relative to the biophysical limits of the ecosystems which are involved as sources or sinks. Ecological Footprint differs only in that it involves the throughput involved in all human activities. Most, but not all, of these activities are economic ones. The Ecological Footprint is connected to many of the other approaches to thinking about and measuring scale:

- Footprint and biocapacity is a way to measure historical human carrying capacity. Most *Carrying Capacity* studies try to answer a hypothetical question: how many people could live on the planet. The Footprint indicates how much of the planet was occupied by people. This is an historical question that can be empirically determined rather than conjecturing on future possibilities.

- Footprint analysis provides a means of assessing the impact of population, affluence (consumption) and technology identified in the *The IPAT Equation*.

- Footprint was used extensively in last update of the *Limits to Growth* to give a summary report of human demand on nature.

- Footprint translates material flows in areas necessary to support these flows.

- Footprint translates some of the principles of *The Natural Step* into a resource account (particularly principle 1 and 3).

- Footprint is an ecological economics tool.

Ecological Footprint could also be useful in making socio-political decisions regarding optimal scale. Optimal scale is an ecological and socio-political target for sustainability. The Footprint accounting process could be used to describe a rough, if cautious, target of optimal scale. The global target would have to be some level of global hectares below those available, to ensure overshoot does not inadvertently occur. By identifying who is contributing how much to the size of a footprint, it can help us understand the potential tra-deoffs in setting optimal scale at different levels. It could also be used to identify throughput targets for various nations, industries, or regions.

*Future Directions*

Efforts are underway to standardize and refine the methodology underlying the Footprint, and to incorporate areas or issues not cur-rently captured.[33] This continuous attention to methodological and conceptual rigor is a positive move and promises to increase the usefulness of this sustainability indicator. The intuitive appeal of the Footprint is another asset, leading to its adoption for many pro-jects. For applications of the Footprint to sustainable scale issues, it would be wise to keep in mind that this measure likely provides an underestimate of ecological impact.

## 5. Critical Natural Capital

*How Much to Use? How Much to Preserve?*

There is no doubt that both using, and preserving, natural capi-tal are essential to human well being. But how much is needed? What aspects are critical? These questions are less easily answered, and engender considerable debate. At one extreme are the economists who argue that financial capital can replace all natu-ral capital (neoclassical economists tend to maintain that man-made capital can, in principle, replace all types of natural capital). At the other extreme are those deep ecologists who argue that no natural capital can be replaced by any other type of capital.

*Weak vs Strong Sustainability*

The concept of weak sustainability holds that all or most forms of natural capital are substitutable by human-derived capital. The

---

33 Global Footprint Network www.footprintnetwork.org

concept of strong sustainability holds that little or no forms of natural capital have human-derived substitutes. Those who believe in weak sustainability are sometimes referred to as "technical optimists," because they believe that technology and human ingenuity will somehow exceed limits imposed by nature. Both logic and the limited information now available support the notion of strong sustainability; technical ingenuity generally serves to increase material and energy throughputs rather than expand the biophysical limits of ecosystems. And it is clear that there are many life supporting ecosystem functions for which there are no substitutes; any substitution that is possible is likely to be marginal.

*The Virtues of Strong Sustainability*

Strong sustainability states that natural and man-made capitals are fundamentally complements rather than substitutes; it states that some natural capital is "critical," in that it consists of assets that are irreplaceable and cannot be substituted by anything else. For example, protection against excessive ultraviolet radiation provided by the atmospheric ozone layer, could theoretically be substituted for by manufactured goods consisting of hats, sun-glasses and suitable clothing. Even if such a substitution would allow human beings to survive, there are no other manufactured goods to prevent the damaging effects on other living creatures, or ecosystem functioning, upon which we depend. The atmospheric ozone layer is an example of "critical natural capital". Similarly, there are other forms of natural capital that face similar substitution difficulties: the global atmosphere, the world's storage capacity or biological diversity, for example.

*Critical Natural Capital: Another Term for Sustainable Scale?*

One answer to the question of how much natural capital is "critical" to human well being is the definition of sustainable scale –

throughput less than regeneration. The concept of critical natural capital (or CNC) is somewhat broader than this definition, and has been the focus of a collaborative international effort to define and operationalize the concept of CNC for application to policy (the CRITINC project). The team has designed a matrix that evaluates different categories of CNC based on seven sustainability principles. These are: maintenance of global environmental processes; protection of biodiversity, critical ecosystems and ecological features; regeneration of renewable resources; prudent use of non-renewable resources; respect for human health standards and critical loads for ecosystems; conservation of landscapes for other human welfare values (aesthetic, spiritual, etc); application of the precautionary principle. It is intended to provide a positive, as opposed to a normative, evaluation of sustainability.

*CRITINC: A Conceptual Approach*

This approach identifies the characteristics of natural capital in the ecosystem being investigated, and relates these characteristics to ecosystem functions that derive from them. An economic Input-Output (I-O) evaluation is then incorporated, describing the impacts of different dimensions of economic activity on ecosystem functions. This gives an estimation of the current status of different types of natural capital, in terms of quantity and/or quality. This assessment is then compared to the sustainability standards mentioned above, expressed as either state or pressure indicators. A state indicator describes the minimum quantity of natural capital necessary for continued functioning, while a pressure indicator explains the maximum pressure that the natural capital stock can tolerate and still maintain its functions. The difference between the actual and sustainable levels is termed the "Sustainability Gap" (SGAP), and provides targets for policy. Evaluation of the least cost method currently available to bring the SGAP to zero through abatement, avoidance or restoration of CNC is also useful for policy design. This can be used to weigh tradeoffs between policy options and evaluate their effectiveness over time. The CRITINC framework is comprehensive in a similar sense to current systems of national accounting. As a composite measure, it can be used to direct policy for overall environmental sustainability, and to weigh the benefits of alternative policy choices that may be made at the expense of environmental sustainability. It can be used to direct policy on a sectoral or regional level by identifying specific sources and results of natural capital depletion.

*CRITINC: An Empirical Approach*

In a study of water resources in the Brittany region of France, for example, CRITINC investigators used this method to identify the role of water-related CNC both ecologically and in the particular societal context of Brittany. This was used to then draft various alternative scenarios that considered uncertainties and incorporated different interests and demands in the local context. This is an important application of the concept of CNC, as it links various indicators to concrete measures used to negotiate between many demands, interests and stakeholders. Beyond determining where a critical level is or would be breached, the framework incorporates economic and social variables to suggest alternative policy choices relevant to local conditions. The framework has been comprehensively applied at local and regional levels, and is also designed for application at the national level. In some countries adequate data now exists on physical inputs and outputs of economic activity to conduct such an analysis. In 1993 the UN Statistical Office advocated the development of physical I-O tables by individual nations, and countries such as Denmark, Germany and the UK have compiled physical I-O tables that are compatible for use with the CRITINC framework (Ekins et al, 2003). The existence of adequate sustainability indicator and I-O data would enable the application of this tool to national policy design in the future. However, use of CNC accounting and analysis is generally lacking at a government level. In Canada for example, natural capital is given recognition as "a critical foundation of our economy" in the 2004 Greening the Budget Submission; however, the assessment only implicitly refers to CNC. The report makes useful suggestions such as improving the availability of environmental information and incorporating natural capital into national accounts. However, its recommendations lack components necessary to link this information with causes and effects related to the economy. The application of the CNC framework, on the other hand, can serve to operationalize ideas and indicators of sustainability, through the application of a structural modeling approach.

*Strengths of the Critical Natural Capital Approach*

Thinking about sustainable scale in terms of the critical natural capital framework developed by CRITINC has demonstrated its usefulness at some regional and locals levels, although applications to date are somewhat limited. It incorporates concepts of both biophysical limits of ecosystems, and the socio-political issues of ethics and social values. It also has the advantage of connecting with national input/output accounting systems now being developed by some nations concerned with the sustainability of their own economies. CNC analysis allows for a clearly defined estimation of the environmental impact of human economic activity relative to sustainable scale. Within this framework sustainable scale is a baseline set by sustainability principles, defined as the minimum quantity and quality of CNC needed to perform critical ecosystem functions.

It is both a measure of overshoot and a measure of what level is sustainable. This target can then be used to monitor change over time, assessing whether human economies are closer or further from a sustainable state. This framework maps out what natural capital is critically important to humans, gives a method for estimating the discrepancy between what is needed and where we are, and translates this to policy application by assigning cost where possible, of reducing that gap to zero. It provides a direction for policy on environmental sustainability, but also a way to measure and recognize tradeoffs between environmental sustainability and other policy goals, where this is the case.

*Limitations of the Approach*

The major limitation in applying this approach, and testing more aspects of its conceptual framework, involve:

- the lack of data regarding economic input/out tables for all nations.
- the lack of information about ecosystem functions and limits.
- the socio-political challenge of identifying what aspects of natural capital are critical from ethical and social perspectives.

These limitations are not unique to the CRITINC approach; the conceptual clarity of the approach which allows it to be empirically examined is a real strength. The lack of data makes empirical examinations of global CNC a continuing challenge.

*Relation to Sustainable Scale*

The concept of critical natural capital identifies a level of sustainable scale, and relates the actual scale of economic activity to this level. By providing an empirical approach, the work of the CRITINC group provides a technique for measuring sustainable scale where the data are available, and contributing to policy development to achieve sustainable scale. This approach can also be used to explore optimal scale. The framework begins from the premise that a certain level of CNC is necessary to support the economy and other aspects of human health and well-being. An optimal scale of economic activity can then be evaluated in terms of support for human health and welfare in a cost-effective manner, after satisfying the requirements for ecological sustainability

## 6. Material Flow Analysis

*What is Material Flow Analysis?*

Material Flow Analysis (MFA) is a quantitative procedure for determining the flow of materials and energy through the economy. It uses Input/Output methodologies, including both material and economic information. It is an accounting system that captures the mass balances in an economy, where inputs (extractions+imports) equal outputs (consumptions+exports+accumulation+wastes), and thus is based on the laws of *Thermodynamics*. Material Flow Analysis recognizes that *Material Throughput* is required for all economic activities and asks whether the flow of materials is sustainable in terms of the environmental burden it creates. It accounts for all materials and energy used in production and consumption, including the hidden flows, or ecological rucksack, of materials that were extracted in the production cycle but which never entered the final products. The physical size of these hidden flows is often many times larger than the flows that end up in actual products. The identification of wastes is a major issue in MFA, as the purpose of conducting a MFA is to minimize the flow of materials while maximizing the human welfare generated by the flow. Its methodology allows for the monitoring of wastes that are typically unaccounted for in traditional economic analyses. As such, it is a method for evaluating the efficiency of using material resources. MFA was developed in Europe, largely at the Wuppertal Institute in Germany, and has been adopted as a methodology by the European Union with respect to its sustainable development program. To date the focus of MFA has been primarily regional or national; a variety of MFA studies have been conducted for both developed countries and economies in transition, including Germany, the UK, Japan, Brazil, Venezuela, Chile and China. EU wide studies have also been conducted, and a global MFA is now underway.

*Strengths of MFA*

MFA provides a direct quantitative measure of the actual material and energy flow through an economy. It quantifies the linkage of environmental problems and human activities, and serves as a systems-wide diagnostic procedure related to environmental problems, supports the planning of adequate management measures and provides for monitoring the efficacy of those measures. MFA allows early warning and supports precautionary measures. It detects problem shifting between regions and sectors. MFA provides aggregated information to support decision making. It can be applied at different levels of economic activity. The procedure makes use of *life cycle analysis* to ensure that all material flows are accounted for. Its use is associated with a variety of sustainable business development practices such as Zero Waste, Increased Resource Productivity, and Extended Producer Responsibility, indicating that it can offer practical, market based solutions to environmental problems. Its emphasis on the efficiency with which material flows contribute to human welfare is a welcomed redirection of the notion of efficiency away from strictly financial flows.

*Limitations of MFA*

Efforts to improve MFA are focusing on ensuring adequate data are available, and that there is more standardization of the methodologies across applications.

*Relation to Sustainable Scale*

MFA is a method for operationalizing the concept of material throughput, and as such is an important contribution to measuring one component of sustainable scale. MFA has proven very useful for demonstrating the negative impacts associated with various specific economic activities. It has highlighted in a very practical way the unsustainability of specific throughputs (e.g. high reliance on non-renewable resources), the incompatibility of absolute levels of input with an equal *Earth-share* identified within the *Ecological Footprint* (e.g. Barrett et al 2002), the displacement of ecological burden to trading partners (generally from developed to undeveloped nations), and the enormous wastes involved in specific products or services.Continued development of this important tool will likely increase its usefulness in linking economic throughput with ecological limits, especially as it is applied to the global level.

## 7. Millennium Ecosystem Assessment

*What is the Millennium Ecosystem Assessment?*

The *Millennium Ecosystem Assessment* (MA) is a United Nations project designed to assess the consequences of ecosystem changes for human well-being. It was initiated in 2001 as the result of an earlier study, *"People and Ecosystems: the Fraying Web of Life"* which reported in 2000 that the world's major ecosystems were in decline, and that significant information about ecosystem functioning was simply not available. The MA was conducted under the auspices of the UNEP, and was governed by a multistakeholder board including international institutions, governments, business, NGOs, and indigenous peoples. Over 1300 experts from 95 countries were involved in producing what is the most comprehensive review of the planet's ecosystems. The objective of the multiyear exercise was to both assess the consequences of ecosystem changes for human well-being, and to establish a scientific basis for action to conserve the sustainable use of ecosystems and their contribution to human well-being.

The MA addressed a series of key questions:

- how have ecosystems and their services changes?
- what has caused these changes?
- how have these changes affected human well-being?
- how might ecosystems change in the future and what are the implications for human well-being?
- what options exist to enhance the conservation of ecosystems and their services to human well-being?

The MA was not designed as a measure of sustainable scale; however, its findings are highly relevant to this issue. The MA Synthesis Report was released in March, 2005.[34]

*Major Findings of the MA*

Some of the key findings of the MA Synthesis Report are perhaps best stated in the report's own words:

*"At the heart of this report is a stark warning. Human activity is putting such strain on the natural functions of Earth that the ability of the planet's ecosystems to sustain future generations can no longer be taken for granted"...*

*"As human demands increase in coming decades, these systems will face even greater pressures - and the risk of further weakening the natural infrastructure on which all societies depend"...*

*"Protecting and improving our future well-being requires wiser and less destructive use of natural assets. This in turn involves major changes in the way we make and implement decisions.*

*Above all, protection of these assets can no longer be seen as an optional extra, to be considered once more pressing concerns such as wealth creation or national security have been dealt with."*

---

[34] Millenium Ecosystem Assessment. "Synthesis Report" (2005) 16.

These observations from the report are based on the following major findings from the Synthesis Report.

*Four Main Findings:*

- over the past 50 years, humans have changed ecosystems more rapidly and extensively than in any comparable period of time in human history, largely to meet rapidly growing demands for food, freshwater, timber, fiber and fuel. This has resulted in a substantial and largely irreversible loss in the diversity of life on Earth.

- the changes that have been made to ecosystems have contributed to substantial net gains in human well-being and economic development, but these gains have been achieved at growing costs in the form of the degradation of many ecosystem services, increased risks of nonlinear changes, and the exacerbation of poverty for some groups of people. These problems, unless addressed, will substantially diminish the benefits that future generations obtain from ecosystems.

- the degradation of ecosystem services could grow significantly worse during the first half of this century and is a barrier to achieving the Millennium Development Goals.

- the challenge of reversing the degradation of ecosystems while meeting increasing demands for their services can be partially met under some scenarios that the MA has considered but these involve significant changes in policies, institutions and practices, that are not currently under way. Many options exist to conserve or enhance specific ecosystem services in ways that reduce negative tradeoffs or that provide positive synergies with other ecosystem services.

The MA provides a detailed summary of the many ways in which human use of ecosystems is unsustainable, getting worse, and posing serious threats to human societies. A popularized, but scientifically accurate, summary of the full, lengthy report is available[35].

*Strengths of the MA*

The MA is one of the best descriptions of the impacts of human activities on ecosystems around the world, and has several strengths:

- there is an explicit connection between ecosystem functions and human well-being; the disproportionate impact of ecosystem degradation on the poor is acknowledged, while clearly stating that wealthy nations will also be affected.

- there is recognition that the situation regarding ecosystem degradation is serious and getting worse; the report clearly states that use of many ecosystems is currently unsustainable.

- there is recognition that there will be increasing demands on these already strained ecosystems if we continue our present course of activities.

- there is recognition of the potential for abrupt, non-linear changes in ecosystem functioning as a result of continuing strains on these systems, and that the timing of these non-linear changes are difficult to predict.

- there is explicit recognition that the non-market functions of ecosystems have great value in terms of human well-being and that market mechanisms alone are not able to conserve and restore ecosystem functioning.

- there is explicit recognition that major changes in economic activities, business operations, institutional and government decision making and life-style adjustments are all needed if ecosystem services are to be sustainable; economic growth is identified as

---

[35] Scientific Facts on Ecosystem Change: Summary of Synthesis Report by Millenium Ecosystem Assessment. Green Facts. www.GreenFacts.org/ecosystems

one of the drivers of ecosystem decline (but not as the main driver).

- the report identifies a variety of mechanisms to restore and conserve ecosystem services; furthermore, it acknowledges that most of the needed changes are not currently being implemented. Some of the needed solutions are identified below.

*What Can we do about it?*

Some key steps available to reduce the degradation of ecosystem services:

*Change the economic background to decision-making*

- make sure the value of all ecosystem services, not just those bought and sold in the market, are taken into account when making decisions.
- remove subsidies to agriculture, fisheries and energy that cause harm to people and the environment.
- introduce payments to landowners in return for managing their lands in ways that protect ecosystem services, such as water quality and carbon storage, that are of value to society.
- establish market mechanisms to reduce nutrient releases and carbon emissions in the most cost-effective way.

*Improve policy, planning, and management*

- integrate decision-making between different departments and sectors, as well as international institutions, to ensure that policies are focused on protection of ecosystems.
- include sound management of ecosystem services in all regional planning decisions and in the poverty reduction strategies being prepared by many developing countries.
- empower marginalized groups to influence decisions affecting ecosystem services, and recognize in law the local communities' ownership over natural resources.

- establish additional protected areas, particularly in marine systems, and provide greater financial and management support to those that already exist.
- use all relevant forms of knowledge and information about ecosystems in decision-making, including the knowledge of local and indigenous groups.

*Influence individual behavior*

- provide public education on why and how to reduce consumption of threatened ecosystem services.
- establish reliable certification systems to give people the choice to buy sustainably harvested products.
- give people access to information about ecosystems and decisions affecting their services.

*Develop and use environment-friendly technology*

- invest in agricultural science and technology aimed at increasing food production with minimal harmful trade-offs.
- restore degraded ecosystems.
- promote technologies to increase energy efficiency and reduce greenhouse gas emissions.

This is an important report and contains much useful information about ecosystem services and their connections to human well being. It adds to our knowledge of ecosystem services (while acknowledging there continues to be serious gaps in what we know). As a consensus report of over 1300 international experts, and considering its sponsorship by the United Nations General Assembly, it is a major wake-up call to all nations and all peoples of the grave situation our planet is in.

*Limitations of the MA*

Strengths can also be weaknesses. As a consensus of multiple stakeholders the report had to be acceptable to all parties involved, including the governments who make up the United Nations, and those with vested interests in the status quo. One of the things that is remarkable about the report is the strength of its conclusions despite the need to satisfy all participant groups, and to not find fault with governments or other sectors who have contributed, knowingly or otherwise, to the ecological degradation the report summarizes. While the report does add detail to what we know about ecosystem services, in a sense there is nothing new in the report in that it has been know for some time that many ecosystems are being used unsustainably. What the report lacks is an explicit statement that the solution to the problem is not a matter of more scientific information, but of political will. The report identifies a variety of causes of ecosystem degradation, but fails to articulate the underlying, common element in all the identified sources, namely material throughput driven by economic growth. The MA conceptual framework does not explicitly use a sustainable scale perspective, but is nonetheless compatible with such an approach. Much of the content of the report could be reframed within a sustainable scale perspective, and many of the solutions proposed are similar to those proposed here. A sustainable scale perspective can help place the MA findings into a broader conceptual framework; such a step would assist in assessing the risks involved in further ecosystem degradation, and point the way to additional solutions.

## 8. Panarchy

*What is Panarchy?*

Panarchy is a conceptual framework to account for the dual, and seemingly contradictory, characteristics of all complex systems – stability and change. It is the study of how economic growth and human development depend on ecosystems and institutions, and how they interact. It is an integrative framework, bringing together ecological, economic and social models of change and stability, to account for the complex interactions among both these different areas, and different scale levels. Panarchy's focus is on management of regional ecosystems, defined in terms of catchments, but it deals with the impact of lower, smaller, faster changing scale levels, as well as the larger, slower supra-regional and global levels. Its goal is to develop the simplest conceptual framework necessary to describe the twin dynamics of change and stability across both disciplines and scale levels. The development of the panarchy framework evolved out of experiences where "expert" attempts to manage regional ecosystems often resulted in considerable degradation of those ecosystems (Gunderson and Holling, 2002). Regional management efforts are generally linear in nature, targeting the maintenance of certain variables – forest growth rates, river clarity, fish harvest rates, etc. It was noted that focusing on managing a single variable, usually one of economic interest, generally resulted in other variables in the system changing, sometimes abruptly, and eventually degrading the entire ecosystem. It was also noted that the changes triggered by attempting to sustain a particular variable were changes that occurred so slowly (over decades or more), that they often went unnoticed until they in turn triggered an abrupt change (e.g. the forest became infested, the river became polluted, or the fish stock collapsed).

*Basic Concepts in Panarchy - Ecosystem Characteristics*

Empirical evidence of natural, disturbed and managed ecosystems identifies four key characteristics:

- change is neither continuous and gradual, nor continuously chaotic. It is episodic, regulated by interactions between fast and slow variables.

- different scale levels concentrate resources and potential in different ways, and non-linear processes reorganize resources across levels.
- ecosystems do not have a single equilibrium; multiple equilibria are common. Ecosystems have processes that maintain stability in terms of productivity and biogeochemical cycles; as well as processes that are destabilizing, which provide diversity, resilience and opportunity.
- management systems must take into account these dynamic features of ecosystems and be flexible, adaptive and experiment at scale levels compatible with the levels of critical ecosystem functions.

*Stages of the Adaptive Cycle: Basic Ecosystem Dynamics*

Panarchy identifies four basic stages of ecosystems: exploitation, conservation, release and reorganization. All ecosystems, from the cellular to the global level, are said to go through these four stages of a dynamic adaptive cycle:

- *exploitation* stage is one of rapid expansion, as when a population finds a fertile niche in which to grow.
- *conservation* stage is one in which slow accumulation and storage of energy and material is emphasized as when a population reaches carrying capacity and stabilizes for a time.
- *release* occurs rapidly, as when a population declines due to a competitor, or changed conditions.
- *reorganization* can also occur rapidly, as when certain members of the population are selected for their ability to survive despite the competitor or changed conditions that triggered the release.

*Adaptive Cycles*

The four stages of the adaptive cycle described above (analogous to birth, growth and maturation, death and renewal), have three properties that determine the dynamic characteristics of each cycle:

- *potential* sets the limits to what is possible - the number and kinds of future options available (e.g. high levels of biodiversity provide more future options than low levels).

- *connectedness* determines the degree to which a system can control its own destiny through internal controls, as distinct from being influenced by external variables (e.g. temperature regulation in warm blooded animals, which involves five different physiological mechanisms, is an example of high connectedness).

- *resilience* determines how vulnerable a system is to unexpected disturbances and surprises that can exceed or break that control (see below for more details).

The adaptive cycle is the process that accounts for both the stability and change in complex systems. It periodically generates variability and novelty, either internally such as through genetic mutations or adaptation, or by accumulating resources that change the internal dynamics of an ecosystem. These changes are the triggers for experimentation. In the reorganization stage various experiments are tested and resources are reorganized in new configurations, some of which enter a new exploitation stage to repeat the cycle.

*Interconnectedness of Levels*

Panarchy places great emphasis on the interconnectedness of levels, between the smallest and the largest, and the fastest and slowest. The large, slow cycles set the conditions for the smaller, faster cycles to operate. But the small, fast cycles can also have an impact on the larger, slower cycles. There are many possible points of interconnectedness between adjacent levels; however, two specific points are of particular interest with respect to sustainability:

- *Revolt* : this occurs when fast, small events overwhelm large, slow ones, as when a small fire in a forest spreads to the crowns of trees, then to another patch, and eventually the entire forest.
- *Remember* : this occurs when the potential accumulated and stored in the larger, slow levels influences the reorganization. For example, after a forest fire the processes and resources accumulated at a larger level slow the leakage of nutrients, and options for renewal draw from the seed bank, physical structures and surrounding species that form a biotic legacy.

The fast levels invent, experiment and test; the slower levels stabilize and conserve accumulated memory of past, successful experiments. Sustainability in this framework is the capacity to create, test and maintain adaptive capability. Development becomes the process of creating, testing and maintaining opportunity.

*Resilience*

Resilience is the capacity of an ecosystem to tolerate disturbances without collapsing into a qualitatively different state. The greater the resilience is in a particular ecosystem the more it can resist large or prolonged disturbances. If resilience is low or weakened, then smaller or briefer disturbances can push the ecosystem into a different state, where its dynamics change. According to this model, after a disturbance, ecosystems evolve through time as ecological niches fill in (increasing connectedness), biomass accumulates (increasing potential) and more successful species outcompete less successful species (decreasing resilience). This makes ecosystems vulnerable to exogamous shocks that they cause a release of resources and a period of rapid reorganization. Once resilience is overwhelmed and an ecosystem enters a new state, restoration can be complex, expensive, and sometimes even impossible. Research suggests that to restore some systems to their previous state requires a return to environmental conditions well before the collapse. Resilience can be degraded by a large variety of factors which largely depend on underlying, slowly changing variables such as climate, land use, nutrient stocks, human values and policies. Resilience is a characteristic of natural systems. When resilience is weakened it is sometimes possible to restore it. Diversity is believed to be a key issue in restoring resilience – both biological and social diversity are important to the extent they contribute functional redundancy (i.e. similar services can be provided by some element in the diversity). But as biological diversity is lost, or as human systems and institutions become homogenous and rigid, then the likelihood of restoring lost resilience declines. The ability to anticipate and plan for the future is a unique characteristic of human systems, and has the potential to increase their resilience.

*Strengths of Panarchy*

Panarchy is a complex and controversial framework for describing ecosystem and human system dynamics and interactions, and it is beyond the scope of this overview to provide a thorough critique. Despite its broad sweep it does have the advantage of relative simplicity in terms of the basic concepts used to describe an array of complex phenomena. This framework developed over several years, is solidly based in empirical research across a broad range of ecosystems, and continues to develop conceptually and generate policy relevant research. Panarchy is a sophisticated attempt to connect ecosystem functioning with economic activities and human institutions for managing the relation between the two. It is an evidence-based approach that forces us to think in non-linear terms about complex systems, while providing the conceptual tools to understand the complexities involved.

*Limitations*

Panarchy remains a hypothesis, despite the many empirical studies it has generated. It's broad sweep requires more empirical testing. While it pretend to be an integrative model of ecological, economic and social dynamics, its focus is primarily ecological. There are competing attempts at integration,[36] which may also account for the observed phenomena. There are also different ways of thinking about resilience (e.g. Fraser et al, in press). Despite these limitations, the panarchy framework continues to stimulate constructive debate and guide empirical studies.

*Relation to Sustainable Scale*

Many of the conclusions and observations made within the Panarchy framework are congruent with those regarding sustainable scale. There is recognition that:

- due to the inherent instability of ecosystems it is extremely difficult to detect or predict transitions to new ecosystem equilibria (e.g. when maximum scale might occur).
- sustainability is about retaining capabilities to continue contributing ecosystem services (i.e. natural income does not deplete natural capital).
- resilience, the ability to resist disturbances, is a key characteristic of ecosystems (e.g. when throughput exceeds regeneration, resilience is reduced).
- uncertainty is an inherent characteristic of the adaptive cycle and must be a key factor in any ecosystem management activity.

---

[36] Fraser, E. "Social vulnerability and ecological fragility: building bridges between social and natural sciences using the Irish potato famine as a case study," *Conservation Ecology*, 7.2 (2003): 9.

- both uncertainty and risk increases with scale (i.e. problems at the global scale pose the greatest risks).
- precautionary policies are necessary to limit surprises (surprises increase as more natural income is used than is regenerated.
- the interaction of different time cycles is important, and that by the time efforts to keep fast variables within desired limits (e.g. GHG emissions) are recognized, it may be too late to avoid a major system change (e.g. climate stability).
- science uses uncertainty to drive inquiry, while vested interests use and foster uncertainty to maintain the status quo.
- biodiversity is an important component of resilience, and is therefore important even if the types of biodiversity have no market value.
- new institutions are needed that gather better information on the slow variables, place greater emphasis on the future, maintain social flexibility for adaptive response, and which maintains and restores ecosystem resilience.
- economic globalization contributes to simplification of ecosystems (as well as to their degradation), reducing resilience.

Panarchy is not a way of measuring sustainable scale. It does provide some interesting and challenging conceptual tools to assist in our understanding of how ecosystems and economic activities and institutions interact. It also identifies a variety of practical approaches to restore and conserve ecological sustainability.

## 9. The Natural Step

*The Natural Step: What is It?*

The *Natural Step* (TNS) is not a measure of scale but rather a process or program to achieve sustainable scale. It was devised by a Swedish physician, Dr. Karl Hendrik Robert, who observed that many of his pediatric patients were developing cancers with no family history of the disease. He traced the causes to environmental contaminants, which led him to further investigate the relationship between environmental issues and human health. He engaged with a number of scientists to identify the basic principles or conditions required for sustainability. In 1989 he founded TNS as a not for profit organization to apply these principles to the systemic causes of environmental problems. TNS recognizes that the life support services provided by ecosystems are in decline at the same time as increasing demands are being placed on them by a growing global economy and population. It envisions the process of achieving sustainability as "expanding the funnel" presently constricted by decreasing resources and increasing demands. This expansion occurs by creatively applying the core conditions for sustainability.

*Conditions for Sustainability*

TNS identifies four basic conditions or principles to achieve sustainability. These are:

- substances from the earth's crust must not accumulate in the earth's biosphere.
- substances produced by society must not increase in the biosphere.
- nature's functions and diversity are not systematically impoverished by physical displacement, over-harvesting or other forms of ecosystem manipulation.
- resources must be used fairly and efficiently in order to meet basic human needs globally.

These conditions arise from consideration of the flows of materials from the earth, through human activities and back to the earth. It focuses on the basic science which identifies the natural balances of these flows without human interference, and the impact of human activities which upsets these balances. TNS articulates a general rule, which states that "human activities must not cause deviations from the natural balances that are large in comparison to natural fluctuations. In particular, any deviations should not be allowed to increase systematically."

*Applications of The Natural Step*

TNS has been accepted as an approach to sustainability in a variety of governmental and organizational setting, including both public and private sector operations. TNS operates in several countries, including Sweden, the UK, the USA, Australia, Israel, New Zealand, South Africa, Canada and Japan. It has consulted to several municipalities and corporations, and provides a variety of publications describing its work.[37]

*Relation to Sustainable Scale*

TNS has much in common with the approach to sustainable scale from an ecological economics perspective. Both approaches acknowledge:

- the importance of the natural sciences and focus on the impact of material throughput on ecosystem services.
- the biophysical limits of ecosystems and note how human activities upset the natural balances established through evolutionary processes.
- that upsetting the natural balances results in degradation of ecosystem services.
- that these imbalances lead to a decline in ecosystem services which provide life support services upon which human societies depend.
- that market and non-market solutions are needed to achieve sustainability.

---

[37] www.naturalstep.org

91

- that social justice in terms of meeting basic human needs is essential for sustainability.

TNS provides a simplified conceptual framework and practical application of a sustainable scale perspective. A point of differentiation is the emphasis the two approaches place on economic activities: sustainable scale focuses on economic growth as a major driver of the imbalances between throughput demands and ecosystem limits; TNS does not explicitly focus on economic activities but certainly includes economic activities in its analysis of drivers which cause the imbalances.

TNS has been successful in attracting interest in these important issues at the level of individual cities, states, corporations and other organizations, and in demonstrating its usefulness in these settings. The pity is that larger political units such as national governments or international bodies such as the OECD or the United Nations have not explored this approach.

# Moral and Spiritual Approaches

Spirituality is understood as the vision of what the human being can achieve if he/she fully develops his/her potentials: spirituality refers to the "deepest values and meanings by which people seek to live."[38]

But, what religions have left us with regards "to guarding the world".

*Sustainable Scale Compatible with Spiritual Traditions*

The concept of sustainable scale is not explicitly addressed in any of the world's major spiritual traditions. However, concern for the issues addressed by sustainable scale are shared by the world's spiritual communities. Issues addressed by optimal scale are very directly addressed by the world's major religions, reinforcing the importance of sustainable scale from a moral and spiritual perspective.

*A Common Theme*

A common element of the world's religious and spiritual traditions is a concern for the relationship of humankind to the cosmos. An important theme in these considerations is the relationship between humankind and nature. Different traditions have approached this relationship in very different ways. Ancient Jewish and early Christian teachings placed nature in a subordinate position to humankind. Humans were considered to be above nature in the divine order, and nature was thought to exist for the benefit and exploitation of humans.

Genesis, chapter 1 reads:

---

[38] https://www.hapres.com/htmls/JSR_1095_Detail.html

*"...fill the earth and subdue it; and have dominion over the fish of the seas and over the birds of the air and over every living thing that moves upon the earth..."*

In Genesis chapter 2, the terms are expanded:

*"...as I gave you the green plants, I give you everything...into your hands are they delivered."*


### Dominant Worldview Spreads Globally

The Judeo-Christian perspective was the predominant worldview in Europe for centuries. The advent of the scientific method and the rise of technology in European culture reinforced this worldview, providing practical means to use nature in unprecedented ways. As European colonialism spread so did this worldview. The development of technology and the rise of capitalism reinforced this worldview, which has now become dominant across the globe.

### A Minority But Consistent Perspective

However, within the Judeo-Christian tradition there has been a consistent minority perspective that contrasts with the notion of nature existing for the purpose of human exploitation. The 10th century Jewish philosopher and theologian Maimonides wrote:

*"It should not be believed that all beings exist for the sake of the existence of man. On the contrary, all other beings, too, have been intended for their own sakes, and not for the sake of something else."*

Saint Francis of Assisi taught that all creatures were equal parts of creation, and not created simply for the pleasure of humankind.

### A Growing Interest in Nature

In 2002, a joint declaration by Pope John Paul II and the Orthodox Ecumenical Patriarch Bartholomew emphasized that humankind is entrusted to guard and protect all creation. Similar statements have recently been made by leaders in the Hindu, Buddhist and Jewish faiths, as well as by Muslim, Baha'i and indigenous leaders, all emphasizing the interdependence of humankind and the natural world. Several interfaith groups have made similar declarations.

*The Christians*

*In 2020 Pope Francis[39] "pulls no punches"* when lamenting pollution, climate change, a lack of clean water, loss of biodiversity, and an overall decline in human life and a breakdown of society.[40] *"Never have we so hurt and mistreated our common home as we have in the last two hundred years"*, he states.[41]

Still, *"describe a relentless exploitation and destruction of the environment, for which he blamed apathy, the reckless pursuit of profits, excessive faith in technology and political shortsightedness".*[42] *Laudato si'* "unambiguously accepts the scientific consensus that changes in the climate are largely man-made"[43] and states that *"climate change is a global problem with grave implications: environmental, social, economic, political and for the distribution of goods. It represents one of the principal challenges facing humanity in our day"* and warns of *"unprecedented destruction of ecosystems, with serious consequence for all of us"* if prompt climate change mitigation efforts are not undertaken.

The encyclical highlights the role of fossil fuels in causing climate change. *"We know that technology based on the use of highly polluting fossil*

---

[39] https://en.wikipedia.org/wiki/Laudato_si%27#Environmentalism

[40] https://www.osv.com/Shop/PDFs/P1747_web_watermark.pdf

[41] http://w2.vatican.va/content/francesco/en/encyclicals/documents/papa-francesco_20150524_enciclica-laudato-si.html

[42] https://www.nytimes.com/2015/06/19/world/europe/pope-francis-in-sweeping-encyclical-calls-for-swift-action-on-climate-change.html

[43] https://cruxnow.com/church/2015/06/popes-environmental-manifesto-looks-like-a-game-changer-in-the-us/

*fuels – especially coal, but also oil and, to a lesser degree, gas – needs to be progressively replaced without delay"*, Francis says. *"Until greater progress is made in developing widely accessible sources of renewable energy, it is legitimate to choose the less harmful alternative or to find short-term solutions."* The encyclical's comments on climate change are consistent with the scientific consensus on climate change.[44]

Concerning the modern technology, the *"dominant technological paradigm"* is seen as a key contributor to the environmental crisis and human suffering. While the technocratic paradigm (i.e. the simulation) is switched on, Pope Francis points out, technology is viewed as *"principal key to the meaning of existence"* and asks the world to *"resist"* its *"assault"*. *"The technological paradigm has become so dominant that it would be difficult to do without its resources and even more difficult to utilize them without being dominated by their internal logic. It has become countercultural to choose a lifestyle whose goals are even partly independent of technology… Technology tends to absorb everything into its ironclad logic, and those who are surrounded with technology 'know full well that it moves forward in the final analysis neither for profit nor for the well-being of the human race"* (see note 40).

*Islamic Traditions*[45]

The over arching principle in the use of nature is derived from the prophetic declaration that states: *"There shall be no damage and no infliction of damage"*. The right to benefit from the essential environmental elements and resources such as water, minerals, land, forests, fish and wildlife, arable soil, air and sunlight is in Islam, a right held in common by all members of society. Each individual is entitled to benefit from a common resource subject to establishing the degree

[44] https://www.nytimes.com/2015/06/19/science/earth/pope-francis-aligns-himself-with-mainstream-science-on-climate.html

[45] https://www.ecomena.org/islam-sustainable-development/

of need, (needs have to be distinguished from wants) and the impact on the environment. Earth is mentioned 61 times in the Qura'n. According to Islam, the universe has been created by Allah (God) with a specific purpose and for a limited time. The utilization of natural resources (*ni'matullah* – the gifts of Allah) is a sacred trust invested in mankind; he is a mere manager and not an owner, a beneficiary and not a disposer. Side by side, the Islamic nation has been termed as (*ummatan wasatan*) the moderate nation in the Qura'n, a nation that avoids excesses in all things. Thus, Muslims in particular have to utilize the earth responsibly for their benefit, honestly maintain and preserve it, use it considerately and moderately, and pass it on to future generations in an excellent condition. This includes the appreciation of its beauty and handing it over in a way that realizes the worship of Allah. The utilization of all natural resources – land, water, air, fire (energy), forests, oceans – are considered the right and the joint property of the entire humankind. Since Man is *Khalifatullah* (the vicegerent of Allah) on earth, he should take every precaution to ensure the interests and rights of others, and regard his mastery over his allotted piece of land as a joint ownership with the next generation.

About *Land Reclamation* Prophet Muhammad said, *"Whosoever brings dead land to life, for him is a reward in it, and whatever any creature seeking food eats of it shall be reckoned as charity from him"*. The Prophet in another occasion said, *"There is no Muslim who plants a tree or sows a field for a human, bird, or animal eats from it, but it shall be reckoned as charity from him"*; and, *"If anyone plants a tree, no human nor any of the creatures of Allah will eat from it without it being reckoned as charity from him"*. This testifies the importance the Prophet in the early days of Islam has given to reclamation of land and the equal rights of all God's creatures to benefit from the resources of earth.

About *Wildlife Protection* and *Natural Resources*, are protected under Shariah (Rules of Islam) by zoning around areas called *"hima"*. In such places, industrial development, habitation, extensive grazing, are not allowed. The Prophet himself, followed by the Caliphs of Islam, established such *"hima"* zones as public property or common lands managed and protected by public authority for conservation of natural resources.

In the Shariah, for *Water Rights* there is a responsibility placed on upstream farms to be considerate of downstream users. A farm beside a stream is forbidden to monopolize its water. After withholding a reasonable amount of water for his crops, the farmer must release the rest to those downstream. Furthermore, if the water is insufficient for all of the farms along the stream, the needs of the older farms are to be satisfied before the newer farm is permitted to irrigate. This reflects the sustainable utilization of water based on its safe yield.

The rights to benefit from nature are linked to accountability and maintenance or conservation of the resource and the *Environment Protection*. The fundamental legal principle established by the Prophet Muhammad is that *"The benefit of a thing is in return for the liability attached to it."* Much environmental degradation is due to people's ignorance of what their Creator requires of them. People should be made to realize that the conservation of the environment is a religious duty demanded by God. God has said: *"And do good as Allâh has been good to you. And do not seek to cause corruption in the earth. Allâh does not love the corrupters"*, (Al Qasas 28:77). Islam calls for the efficient use of natural resources and *Waste Minimization*. God says in Qura'n: *"Eat and drink, but waste not by excess; "He" loves not the excessive"*, (Al-A'raf 7:31). *"And do not follow the bidding of the excessive, who cause corruption in the earth and do not work good"*, (Ash-Shu'ara 26: 151-

152). *"And do not cause corruption in the earth, when it has been set in order"*, (Al-A'râf 7:56).

*Water Pollution and Conservation* also plays another socio-religious function: cleaning of the body and clothes from all dirt, impurities, and purification so that mankind can be presentable at all times. Only after cleaning with pure (colorless, odorless and tasteless) water, Muslims are allowed to pray. One can only pray at a place that has been cleaned. In light of these facts, Islam stresses on preventing pollution of water resources. Urinating in water (discharging wastewater into water stream) and washing or having a bath in stagnant water are forbidden acts in Islam. The Prophet said: *"No one should bathe in still water, when he is unclean"*. The teachings of Prophet Muhammad emphasize the proper use of water without wasting it. The Prophet said: *"Don't waste water even if you are on a running river"*. He also said: *"Whoever increases (more than three), he does injustice and wrong"*.

Islamic legislation on *Sustainable Forestry* (the preservation of trees and plants) finds its roots in Qura'nic teachings of Prophet. They include the following: *"Whoever plants a tree and looks after it with care, until it matures and becomes productive, will be rewarded in the Hereafter"* and *"If anyone plants a tree or sows a field and men, beasts or birds eat from it, he should consider it as a charity on his part"*. He is also reported to have encouraged tree planting as a constructive practice, saying that even if one hour remained before the final hour and one has a palm-shoot in his hand, he should plant it. Even at times of war, Muslim leaders, such as Abu Baker, advised their troops not to chop down trees and destroy agriculture or kill an animal.

The protection, conservation, and development of the environment and natural resources is a mandatory religious duty to which every Muslim should be committed. This commitment emanates from the individual's responsibility before God to protect himself and his community. God has said, *"Do good, even as God has done you good, and do not pursue corruption in the earth. God does not love corrupters"*.

*Jewish traditions*[46]

Jewish tradition teaches us to care for our planet in order to preserve that which God has created. Psalm 24 notes, *"The earth is the Lord's and the fullness thereof,"* a dramatic assertion of God's ownership of the land. It follows, then, that any act that damages our earth is an offense against the property of God. The Jewish concept of *bal tashchit*, *"do not destroy,"* forbids needless destruction. Judaism emphasizes our need to preserve our natural resources and generate new ones for future generations. The Talmud tells the story of the sage *Choni*, who was walking along a road when he saw a man planting a carob tree. Choni asked, *"How long will it take for this tree to bear fruit?"* *"Seventy years,"* the man replied. Choni then asked, *"Are you so healthy that you expect to live that length of time and eat its fruit?"* The man answered, *"I found a fruitful world because my ancestors planted it for me. Likewise, I am planting for my children."* In fact, tradition values this concept so much that the rabbis teach that if a man is planting a tree and the messiah appears, he should finish planting the tree before going to greet him (*Avot d'Rebbe Natan* 31b).

We are encouraged *l'vadah ul'shamrah*, *"to till and to tend,"* to become the Earth's stewards. In Isaiah 41:17-18, God promises, *"I, the God of Israel, will not forsake them. I will open rivers in high places and fountains in the midst of valleys; I will make the wilderness a pool of water and the dry land springs of water."* In other words, we were given our planet as a loan from God, and we should work to preserve it.

Among the many issues facing our planet, climate change poses a huge challenge to resource development and even daily habits.

---

[46] https://reformjudaism.org/jewish-views-environment

Addressing climate change requires us to learn how to live within the ecological limits of the earth so that we will not compromise the ecological or economic security of those who come after us.

The Torah commands, *"Justice, justice shall you pursue"* (Deuteronomy 16:20), and thus, our energy policy must also be equitable and just - and the countries most responsible for climate change should be those most responsible for finding a solution to the problem. Judaism also underscores the moral imperative of protecting the poor and vulnerable: *"When one loves righteousness and justice, the earth is full of the loving-kindness of the Eternal"* (Psalms 33:5). Indeed, poor nations are likely to bear the brunt of the negative impacts associated with climate change.

Because our sacred texts teach that humankind has an obligation to improve the world for future generations, Jewish tradition encourages families and communities to reduce their waste and make smart consumer choices, investing in companies that do not pollute and supporting behaviors and policies that encourage conservation.

As one of the most important natural resources to humanity's survival, water has a special place in Jewish tradition, playing a role in nearly every major story in the bible. Isaac's wife was chosen for him at a well; the baby Moses was saved after floating down a river; the Israelites were freed when the red sea parted; Miriam will forever be remembering by her gift of water to the Jewish people in the desert. Our clean, fresh water supplies and mineral resources are being exhausted by industrial and population growth, and it is vital that we lead in conservation while developing natural resources. Jewish tradition has long advocated that local and national governments take appropriate measures to remove or ameliorate the growing threats of environmental pollution and to afford protection to the environment.

The principle of *pikuach nefesh*, saving human lives above all else, is our greatest moral obligation. We are taught, *"You shall not stand idly by the blood of your neighbor"* (Leviticus 19:16), and to *"choose life, that you and your descendants may live"* (Deuteronomy 30:20). It follows, then, that Jewish values command us to preserve the earth and its varied life for our sake and for generations to come. It is our obligation to preserve human life by educating ourselves about the dangers of environmental health risks and working to prevent them for the sake of all humanity.

As heirs to a tradition of stewardship that goes back to Genesis and teaches us to be partners in the ongoing work of creation, we cannot accept the escalating destruction of our environment and its effect on human health and livelihood. It is our sacred duty to alleviate environmental degradation and the human suffering it causes instead of despoiling our air, land, and water.

*Jehovah's witnesses traditions* [47]

Human activities may be damaging the health of our planet now more than at any time in history. As the threats of such problems as global warming become more alarming, scientists, governments, and industrial groups are increasing their efforts to respond. Do we as individuals have a responsibility to help care for the environment? If so, to what degree? The Bible provides good reasons for us to act in ways that benefit the earth. It also helps us to be balanced in our efforts.

Jehovah God made the earth to be a garden-like home for mankind. He pronounced all of his work to be *"very good"* and assigned man *"to cultivate (the earth) and to take care of it"* (Genesis 1:28, 31; 2:15). How does God feel about earth's present condition? Clearly, he is deeply offended by man's mismanagement, for Revela-

---

[47] https://wol.jw.org/en/wol/d/r1/lp-e/102007443

tion 11:18 foretells that he will *"bring to ruin those ruining the earth"*. So we should not be indifferent to the earth's plight.

The Bible assures us that every trace of the damage caused by man will be undone when God 'makes all things new' (Revelation 21:5). However, we should not conclude that since God will in time restore the earth, our actions now do not matter. They do! How can we demonstrate that we share God's view of our planet and support his will for it to be a paradise?

Normal human activities produce a measure of waste. Jehovah wisely designed earth's natural cycles to process such waste, cleaning the air, the water, and the ground. (Proverbs 3:19) Our actions should be in harmony with those processes. Thus, we need to be careful not to contribute unnecessarily to earth's environmental woes. Such care shows that we love our neighbor as ourselves (Mark 12:31). Consider an interesting example from Bible times.

God instructed the nation of Israel to bury human waste *"outside the camp"* (Deuteronomy 23:12, 13). This kept the camp sanitary and sped up the process of decomposition. Similarly today, true Christians strive to dispose of garbage and other waste quickly and properly. Special care should be taken when disposing of toxic materials.

Many waste products can be reused or recycled. If recycling is mandated by local laws, then obeying such laws is part of rendering *"Caesar's things to Caesar"* (Matthew 22:21). Recycling may require extra effort, but it demonstrates a desire for a clean earth.

In order to fill our needs for food, shelter, and fuel and thus sustain our lives, we must consume natural resources. How we use those resources reveals whether we recognize that they are gifts from God. When the Israelites desired meat to eat in the wilderness, Jehovah provided an abundance of quail. Greed caused them selfishly to abuse that gift, greatly angering Jehovah God (Numbers 11:31-33). God has not changed since then. Accordingly, responsible Christians avoid needless waste, which could be a sign of greed. Some may view the unlimited consumption of energy or other resources as their right. But natural resources should not be squandered simply because we can afford them or there is an abundance. After Jesus miraculously fed a large crowd, he directed that the remaining fish and bread be gathered (John 6:12). He was careful not to waste what his Father had provided.

Every day we make choices that affect the environment. Must we take an extreme approach, withdrawing from human society to avoid any negative impact on the earth? Nowhere does the Bible recommend such a course. Consider Jesus' example. While on earth, he led a normal life, which allowed him to accomplish his God-assigned preaching work (Luke 4:43). Furthermore, Jesus refused to get involved in politics as a means of solving the social problems of his day. He clearly stated: *"My kingdom is no part of this world"*, John 18:36. It is proper, though, for us to consider the environmental impact of our choices in such areas as household purchases, transportation, and recreation. For example, some choose to purchase products that have been produced or that operate in ways that minimize damage to the environment. Others strive to reduce their share in activities that create pollution or unduly consume natural resources. There is no need for one person to enforce his environmental decisions on others. Personal and local circumstances vary. Still, we remain individually accountable for our decisions. As the Bible states, *"each one will carry his own load"*, Galatians 6:5. The Creator placed upon humans the responsibility to care for the earth. Appreciation for this assignment and humble respect for God and his creative works should motivate us to make thoughtful, conscientious decisions regarding how we treat the earth.

*Eastern Religious Traditions*

In South Asia, the Hindu and Buddhist traditions both encompass the notion of obligations to preserve and protect nature. *"The desire for peace exists everywhere, but the majority of people are not in a position to enjoy peace, stability and security they desire,"* noted venerable Dr Ashin Nyanissara in 2015, spiritual head of the Sitagu International Buddhist University (SIBU), in opening a two-day gathering of spiritual leaders and scholars, in Yangoon, Myanmar.[48] The SAMVAD[49] initiative is driven by Buddhists and Hindus who are keen to exploit commonalities in their spiritual teachings to create a more tolerant, liberal and accommodative world living in harmony with nature rather than seeing it as a resource to exploit. Both Buddhists and Hindus pointed out that many rituals and festivals in their respective religions which have survived so far draw the link between nature and humans.

*"The consciousness that man is part of nature and not independent and certainly not its master is fundamental to protecting and sustaining environment and ecology,"* noted Rajalaksmi Ravi, a social activist from Chennai, in the Indian state of Tamil Nadu. *"Hindu culture has made the tree a symbol of forests and prescribed 'Vriksha Vandana' (reverence of trees) as the attitude of humans to forests – unless humans revere trees, forests are not safe,"* she noted, pointing out that 'Ganga Vandana' (worship of water) and 'Bhumi Vandana' (homage to earth) *"celebrate all rivers, lakes and ponds to inculcate environmental consciousness and protect water resources".* In his

---

[48] https://www.sdgsforall.net/index.php/goal-4/414-eastern-spirituality-could-help-sustainable-development

[49] SAMVAD is an initiative of Indian Prime Minister Narendra Modi and Japanese Prime Minister Shinzo Abe to adopt principles of Asia's age-old spiritual teachings of Hinduism and Buddhism to address modern-day issues threatening human civilisation.

video message, Indian Prime Minister Modi reminded participants that Hindu and Buddhist philosophies see nature as living in harmony. *"If we don't live in harmony with nature, we have climatic change"*, he warned. *"(We must) revere nature and not consider it merely as a resource to exploit"*. *"Buddhists apply the concept of interdependent origination to everything in our world"*, said Tibetan Buddhist monk His Holiness Drikung Kyahgon Chetsang. *"An authentic environmental consciousness will develop naturally once people recognise the deep interdependence between humans, plants, and animals. Thus, the ancient Buddhist philosophy of interdependence is critical to the future of our planet"*, he said. The Tibetan monk described a *"Go Green Go Organic"* campaign his monastic order is developing in the Himalayan Ladakh region of India where water supplies and environment are under threat from global warming. Over 2000 trees have been planted in an effort to prevent soil erosion and also to give local people natural resources to harvest sustainably, which he called *"creating sustainable economic opportunities"*. With the glaciers of the Himalayan snow mountains melting rapidly, his campaign has dug trenches to capture the water flow in the summer and distribute its water to a wider area, which is also giving rise to the growth of wild plants that contribute to tackling soil erosion. *"We need to develop a broader perspective of the earth as a whole"*, argued His Holiness Chetsang: *"Natural disasters and ecological problems do not choose people of one religion or one nation."*

And the Confucian and Taoist traditions of East Asia provide even more dramatic examples of a unity among humankind, nature and all living things.

*Daoism, Confucianism and the environment*

In September 2013[50], an unusual environmental organization was launched in one of the most ancient and significant sites in

---

[50] https://www.chinadialogue.net/culture/6502-Daoism-Confucianism-and-the-environment/en

China - the Songyang Academy, Dengfeng, Henan. Founded in the 11th century AD, this was one of the four Confucian Academies of China. The site was originally Buddhist; it became Daoist in the early 7th century and Confucian in 1035. All of that is but a footnote in history compared with the two ancient pine trees on the site, which were already so venerated in 110 BC that Emperor Wu of the Han Dynasty went there to worship them. It was in the presence of these two great ancient trees that the International Confucian Ecological Alliance (ICEA) was inaugurated. It marked the first time a specific Confucian organizational response has emerged to the environmental issues confronting not just China, but (as ICEA's title says) the whole world. The Confucianists are planning to designate two cities as models of Confucian ecological values: Confucius's birthplace of Qufu in Shandong, and Dengfeng in Henan. Both commitments are part of their membership of the Green Pilgrimage Network (GPN), which links pilgrim places around the world in their journey to become greener.

They worked on developing an eight-year plan of environmental action based around their 500 temples and dozens of academies. They also planned, in association with the Ministry of Education, to sponsor 100 lecture events on Confucianism and ecology, and establish an academic board to oversee scholarly research on Confucian values and ecology. In 2014 they celebrated the 2,565th birthday of Confucius with a huge gathering in Beijing focusing on ecology.

Earlier in 2013, at Asia House in London, the China Daoist Association announced the further development of their 20 years of action on ecology. President of the Association, Master Ren Farong, spoke powerfully about his faith's protection of sacred mountains; its involvement in developing Green Pilgrimage in China and commitment to stop the illegal wildlife trade, through restoring Traditional Chinese Medicine (TCM) to its proper, herbal foundations. All this shows that the desire to find ways of being and living which reflect deep Chinese cultural traditions, wisdom and insights is now a considerable movement not just amongst young Chinese but across all age groups. Daoism and Confucianism, as the two indigenous spiritual and philosophical traditions of China, are at the very core of the recovery of a specifically Chinese perspective on protecting our planet.

When they created their first statement on Daoism and the environment 20 years ago, the Daoists traced four principles.

1. *Follow the Earth*

   The Earth respects Heaven, Heaven abides by the Dao, and the Dao follows the natural course of everything. Humans should help everything grow according to its own way. We should cultivate the way of no-action and let nature be itself. In a survey of the nine major sacred mountains of China (five Daoist ones: four Buddhist) in the late 1990s it was found that where religious communities had managed to return to the sacred mountains after the Cultural Revolution, they had helped to protect a far wider range of biodiversity than on any comparable mountains. From this has come a program of Daoist temples identifying indigenous and threatened species and creating nurseries to sustain their recovery.

2. *Harmony with nature*

   In Daoism, everything is composed of two opposite forces known as Yin and Yang. The two forces are in constant struggle within everything. When they reach harmony, the energy of life is created. Someone who understands this point will not exploit nature, but will treat it well and learn from it. This is the philosophical, spiritual and physiological basis of the program of the Daoists to oppose the use of endangered species – including tigers, elephants and rhinos – in what passes for TCM. Daoists are now teaching that if, by sourcing the ingredients you have disturbed the balance of yin and yang in the world, by harming rare species, or giving pain to other creatures, then the medicine will not work. Indeed it will harm you if you take it.

3. *Too much success*

If the pursuit of development runs counter to the harmony and balance of nature, even if it is of great immediate interest and profit, people should restrain themselves from it. Insatiable human desire will lead to the over-exploitation of natural resources. To be too successful is to be on the path to defeat. The critique of consumerism runs deep within the Daoist ecological program and at one level is symbolized by the Three Sticks Movement. Temples noticed that with increased wealth people were bringing huge bunches of incense sticks to burn at the shrines as if they could influence the gods by their excess. So the Daoists brought in a rule of simplicity. Just three sticks of incense is enough, one for Heaven, one for Earth and one for yourself. In one small message is a big idea on how to live sustainably.

4. *Affluence in bio-diversity*

Daoism has a unique sense of value in that it judges affluence by the number of different species. If all things in the universe grow well, then a society is a community of affluence. If not, this kingdom is on the decline. This view encourages both government and people to take good care of nature.

So often the environmental movement is about instilling a sense of guilt and even fear. It doesn't work. This is why Daoism encourages a positive view of the natural richness of nature as a cause for celebration. At Louguantai in Shaanxi where Lao Zi wrote the core Daoist book, the Dao De Jing, the temple has become the major centre of Daoist ecology. It runs monthly ceremonies to celebrate the cycle of nature. These attract thousands of people from the nearby city of Xi'an, who want to be put back in touch with the rhythm of nature. Confucianism is new to this but its first ever Statement on the Environment in 2013 shows where it is going to put its energy over the next few years: into *"a sustainable harmonious relationship between humans and nature"*. As the Shang Shu, one of the Five Confucian Classics, says so pithily in Chapter 22: *"Just do what is right and proper and then all will be well."*

*Indigenous Traditions*

Indigenous peoples from all continents have traditionally had the most intimate connections to nature in their belief systems and world-view. Respecting and protecting nature played an integral part of the everyday lives of these groups who relied so directly on nature for their survival. The Seventh Generation Principle, for instance, is based on an ancient Iroquois philosophy that the decisions we make today should result in a sustainable world seven generations into the future. The first recorded concepts of the Seventh Generation Principle date back to the writing of The Great Law of Iroquois Confederacy, although the actual date is undetermined, the range of conjectures place its writing anywhere from 1142 to 1500 AD. The Great Law of Iroquois Confederacy formed the political, ceremonial, and social fabric of the Five Nation Confederacy (later Six). The Seventh Generation Principle today is generally referred to in regards to decisions being made about our energy, water, and natural resources, and ensuring those decisions are sustainable for seven generations in the future. But, it can also be applied to relationships - every decision should result in sustainable relationships seven generations in the future.

*A Resurgence of Stewardship*

The current resurgence of the stewardship concept among the world's major spiritual traditions is in direct contrast to the notion that humankind has a duty to subdue and exploit nature – the worldview now dominant. The stewardship concept recognizes the dependence of humankind on nature, and makes explicit our obligations to preserve and protect all creation. Any activities that have the potential to trigger an irrevocable collapse of the ecosystem services that support all life, are clear violations of this obligation. The increasing attention devoted by the world's major spiritual traditions to environmental preservation is a hopeful sign. Values consistent with living within the finite scale provided by ecosystem support services, appears to be reemerging.

# CHAPTER II
## The Importance of Sustainable Scale

Why is the meaning and understanding of the sustainability scale so important?

*Of the Global Scope:* over time scale problems have grown to the point they now involve global ecosystems.

*The Severe Consequences:* ecosystems are now providing fewer services critical to human well being then they did in the past.

*The Situation is Getting Worse:* almost all measures of sustainable scale are indicating rapid decline.

*The Problems are Pervasive:* several major global ecosystems are currently either unsustainable or rapidly approaching unsustainability.

*Time is Short:* many of the affected global ecosystems involve long term cycles that delayed our recognition of the problem. Some problems are still not being addressed.

*Considerable Uncertainty Exists:* we know enough now, even though our understanding is incomplete, to take action.

*Counterintuitive Solutions Are Needed:* The necessary scale relevant solutions are contrary to the dominant policy of economic growth.

### 1. Global Scope

*From Local to Global*

Problems of sustainable scale are unprecedented in the history of human civilization.[51] Previous environmental problems were local or regional. Human activities are now affecting several global ecosystems that co-evolved over millions of years, that provided the conditions for life to evolve on the planet, and that are essential for our continued survival. The very fact that our activities are affecting

---

[51] McNeill, J. R. Something New Under the Sun. New York: W. W. Norton Co., 2000.

such robust and essential life support services should be a major cause for concern.

## 2. Severe Consequences

*Irrevocable Harm*

Sustainable scale problems differ from many other environmental problems in significant ways. Most critical is that sustainable scale problems have the potential for irrevocable harm to vital life support systems, such as climate stability, UV radiation protection and the resilience provided by biodiversity. In the past environmental problems were local and the harms done were largely reversible. However, if maximum scale is ever exceeded then the harm done will be irrevocable, the losses will be permanent and there will be no substitutes for the services lost. Even if this worse case scenario is avoided we run a double risk if we exceed sustainable scale. Firstly, the ecosystems services we rely on are reduced in both quantity and quality. Secondly, these reduced services mean we are more vulnerable to exceeding maximum scale as time passes.

*Destroying the Mechanisms of Survival*

If maximum scale were exceeded then not only would ecosystem services be disrupted, but the very biophysical systems which produce these services would also be destroyed. To use a machine analogy, if maximum scale were exceeded we would not only lose the products made by the machine but we would also destroy the machine that makes the products. Furthermore, our ability to rebuild the machine would be lost.

*Compounding Problems*

Exceeding maximum scale is not the only serious threat. Exceeding sustainable scale is also dangerous. Whenever sustainable scale is exceeded ecosystem problems are compounded; the problems can spread from one global system to another, and fewer ecosystem services are available to meet human needs. Global biogeochemical cycles and ecosystems are by their nature complex and interconnected. They co-evolved over millions of years, intricately combining living and non-living systems. Disrupting any one of these major systems by exceeding sustainable scale will inevitably have profound effects on other global systems, potentially creating an escalating cascade of collapses. The compounding of problems can occur in different ways. There is the impact of disruptions in one global system spreading to other systems. Current alteration of the carbon cycle through the emission of greenhouse gases, for example, is not only affecting global climate patterns. The climate changes are having an impact on the recovery of the atmospheric ozone layer, accelerating biodiversity loss, altering animal and human disease vectors, bleaching coral reefs, and generating more frequent and more intense storms. If these systems are themselves under stress from forces independent of the carbon cycle, as is the case, the magnitude of the disruptions will be enhanced. Under these circumstances the risk of pushing these other areas beyond sustainable scale is also increased.

Additional compounding may occur if human induced ecosystem disruptions trigger positive feedback mechanisms, enhancing the effect. Such occurrences increase the likelihood that sustainable scale will be exceeded. This is the caso of the northern permafrost which is melting by global warming and huge quantities of methane stored in these frozen lands are being freed free. Methane is a powerful greenhouse gas and release of these reserves would contribute to yet more climate change.

*Unprecedented and Potentially Disastrous Impacts*

No prior civilization has ever faced the number and magnitude of sustainable scale problems that confront us today. The success of the human enterprise in terms of population growth and economic consumption now has the potential to irreparably disrupt global ecosystems that took millions of years to evolve in an intricate balance of complex interdependencies that allowed life to develop and humans to thrive. Mistakes made by previous civilizations were limited to their own area of control. Disrupting global ecosystems by continued economic growth will affect the entire planet. There will be no new valleys or mountain areas to retreat to - all will be affected. There will be no opportunity to reset the disrupted ecosystem and try again to reap the many benefits we now take for granted. We have one chance to get it right.

## 3. Getting Worse

*Overall Decline in Ecosystem Services*

There are many examples of local and regional environmental problems getting better, especially in the cities of wealthy northern countries. However, there are many signs that ecosystems around the world are no longer able to provide the same level of services they provided in the past. Many significant problems related to sustainable scale issues are global in nature and continue to get worse.[52]

*Some Examples*

On an annual basis more greenhouse gases are emitted driving climate change; biodiversity loss is increasing dramatically, including loss of both species and vital ecosystems such as mangroves, coral reefs and old growth forest; depletion of non-renewable resources critical to human activities, such as phosphates and petroleum, is increasing; and resources that should be renewable, such as various fisheries and soil fertility, are being depleted beyond the point where they can recover in a meaningful human timeframe.

---

[52] https://wwf.panda.org/knowledge_hub/all_publications/living_planet_index2/

## 4. Pervasive Problems

*Ecosystems at Every Level Affected*

Not just one global ecosystem is under threat. Multiple systems at every level - global, regional and local - are affected. A recent report on the most extensive study of global ecosystems ever undertaken concluded that almost every major ecosystem on the planet is under stress. Each of the areas reviewed in *Areas of Concern - Volume II*, are currently of significant problems. Approaching these areas of concern from a sustainable scale perspective indicates that each could conceivably exceed sustainable scale sometime in the present century. Some of these systems have already been pushed beyond sustainable scale. While these areas of concern are as diverse as oil production and biodiversity loss, there is an underlying commonality - they have a common cause - material throughput as an integral part of the economic process.

*Caution Needed*

The pervasiveness of scale problems adds to both the difficulties in understanding sustainable scale issues and the dangers they pose. It is difficult enough to understand a single complex system operating in a relatively stable natural state. The complexities are increased significantly when multiple complex systems are affected simultaneously, and are interacting in novel and dynamic ways. If we have already driven one or more global systems beyond sustainable scale, as many scientists believe, then the risks of a domino effect triggering yet more systems' changes is increased. Considerable caution is required to ensure all global life support systems are maintained in a sustainable range.

*Pervasiveness of Scale Problems Suggests Focusing on Causes*

The pervasiveness of scale problems indicates that it is not simply a few areas that need attention. Altering a few industrial processes or banning a few toxic compounds will not solve the problem. The causes are more fundamental and widespread. Our current approaches to environmental problems are piecemeal and generally rely on end of pipe solutions, attempting to fix them after they occur. The magnitude and seriousness of sustainable scale problems suggest that focusing on underlying causes would be both more efficient and more effective.

## 5. Considerable Uncertainty

*Uncertainty, Risk and Ignorance*

Uncertainty means we are not sure of what we know. Risk and ignorance are different types of uncertainty. Risk is a type of uncertainty which occurs when we know both the range of outcomes possible, as well as the likelihood of possible outcomes, but not exactly which will occur (as in throwing dice). Pure uncertainty occurs when we know the possible outcomes, but not the likelihood of any particular outcome. And if we do not even know what outcomes are possible, then we are in a condition of ignorance, or absolute uncertainty.

*Certainty Regarding Sustainable Scale*

There is considerable uncertainty regarding many aspects of sustainable scale. But with respect to the impact of any specific level of material throughput, we can be certain that scale outcomes will be either sustainable or unsustainable. Given the biophysical limits of ecosystems, and the second law of thermodynamics, we can also be certain that maximum scale is a possible outcome, as a special case of unsustainable scale. Being aware of it as a possibility means we can take action to avoid it.

*Special Cases of Certainty Regarding Scale*

There are three special cases we can known with certainty regarding sustainable scale:

- we can know with certainty that using non-renewable resources will eventually lead to their depletion. If we know the amount of non-renewable resources available, and the rate of depletion, we can also know with relative certainty, the amount of time until depletion occurs. Estimates of peak oil production are an example.

- we can know with certainty that if we exceed the critical level of harvesting of a renewable resource (called *depensation*[53]), then that resource will eventually collapse. If we know the stock of renewable resource available, the rate of replenishment and the harvest rate, we can also project the conditions under which such a collapse will occur. Such projections are less accurate than with non-renewable resources (as we have learned with fisheries management) because the complexities involved are greater.

- in the case where ecosystem capacity to absorb a substance is zero or near zero, then we can be certain that any throughput of such emissions is unsustainable. Atmospheric ozone depletion provides an example of this phenomena. However, it should be noted that the point where maximum scale might be reached in such special cases remains uncertain.

---

[53] Depensation is when there is a decrease in breeding individuals which leads to reduced production of offspring within a population. This is generally caused by higher levels of predation, the introduction of an invasive species, or the Allee effect (the originator and namesake of the phenomenon was Warder Clyde Allee (1885–1955), a University of Chicago zoologist and animal ecologist, whose special interest was group behavior in animals). Parasitism is another contributor to depensation and a population can completely disappear when affected by depensation. Critical depensation is when levels of depensation are high enough that a population cannot sustain itself and recover. If the population drops below the critical depensation level, a population can collapse and a local extinction is possible. Depensatory effects can be grouped into four distinct mechanisms: reduced fertilization, damaged group dynamics, high concentration of predators, and the surrounding environment. These mechanisms should all be considered when identifying depensatory effects in a population using population dynamics models. Allaby, Michael (2004). Depensation. *A dictionary of Ecology.* Retrieved from http://www.encyclopedia.com/doc/1O14-depensation.html and Liermann, Martin and Hilborn, Ray (2001). Depensation: evidence, models and implications. *Fish and Fisheries:* 2: 33-58. Retrieved from http://www.seaturtle.org/PDF/LiermannM_2001_FishFish.pdf

## Uncertainty Regarding Scale

Determining boundaries for sustainable or maximum scale with any degree of scientific certainty, is at best a long way off. Given the complexities and dynamic interdependencies involved, as well as the many areas of concern, such uncertainties are likely to remain for some time. Indeed, because complex ecosystems are characterized by *Emergent Properties*[54] inherent cycles of growth and decay and non-linear changes under stress, some uncertainty regarding these boundaries is irreducible.

## The Role and Limits of Science

Better scientific understandings of ecosystem functioning and human impacts on such systems are urgently needed to help us assess sustainable scale boundaries. Baseline data for many ecosystems from local to the global levels are seriously wanting. Major scientific efforts to systematically understand major ecosystems are just beginning. Likewise, approaching ecosystems from the perspective of natural capital is also recent. While solid research is needed asking scale relevant questions is also important. But even with a major global effort, completely understanding sustainable scale boundaries is beyond the current ability of science, given the inherent uncertainties involved.

## Managing with Uncertainty

Accepting the reality of uncertainty and the limitations it imposes are critical to managing our predicament. Accepting both the contributions that more scientific research can provide, and the ir-

---

[54] An emergent property is a property which a collection or complex system has, but which the individual members do not have. A failure to realize that a property is emergent, or supervenient, leads to the fallacy of division. Examples of emergent properties include cities, the brain, ant colonies and complex chemical systems. Many emergent properties are the result of interactions between species in communities. For instance, a cow's ability to digest grass depends on bacteria living in the cow's stomach! As communities interact with the physical environment, ecosystems form. Together, the ecosystems on planet Earth form a biosphere.

reducible uncertainties involved, are important in determining how to best manage the situation. Waiting for scientific certainty in determining sustainable scale boundaries will not be fruitful, but dangerous. Given what we know for certain, the stakes involved, and the availability *Attractive Solutions*, the prudent course is to begin managing human activities to ensure we remain within sustainable scale.

### Beyond Science

No amount of scientific information can ever provide complete answers to a key issue in developing scale relevant policies. Even if science could firmly establish the empirical basis of sustainable scale boundaries (however unlikely), policies would have to be designed to ensure an adequate safety margin to account for unanticipated events that could cause an unplanned transgression of those boundaries. Running a global economy on the threshold of maximum sustainable yield, on the sustainability threshold, is hardly a prudent or safe policy.

### Sustainability a Social and Moral (as well as Scientific) Issue

Establishing such a safety margin requires the definition of optimal scale, a social boundary within the boundary of biophysical limits. Optimal scale can be developed despite scientific uncertainty about sustainable scale boundaries. Optimal scale is a socio-political construct requiring extensive consultation concerning a variety of issues such as social justice and responsibility to future generations and other living things. Without some measure of global consensus on this critical issue, there remains the danger of exceeding sustainable scale. There is also the very real danger of some dominant national power or powers, inadvertently or by design, appropriating most available, but ever dwindling, resources for their own survival. Indeed, in addition to the developing countries continuing to appropriate more than their share of the earth's resources based on population, the emerging economies of China, India and other rapidly developing countries are all increasing their competition for dwindling resources.

## 6. Time is Short

*Sustainable Scale Already Exceeded*

Many scientists believe that sustainable scale has already been exceeded in several critical areas. Use of non-renewable resources is by definition not sustainable. As we increase our use of such resources without planning for replacement of the unique benefits they provide, we deprive future generations. Petroleum is a good example. The accelerated rate of biodiversity loss is many times the pre-industrial level. The increasing loss of keystone species means this loss will continue to get worse. We have exceeded the sustainable scale of various fisheries. These resources should be renewable. Our emissions of novel man-made toxic substances, for which ecosystems have little, or no, tolerance is increasing. The production and emission of ozone depleting compounds into the atmosphere quickly exceeded sustainable scale. Concentrations of greenhouse gases are in excess of those in the preindustrial period by over 30 %.

*Effects of Exceeding Sustainable Scale*

Exceeding sustainable scale means that we are living off natural capital rather than natural income. This situation inevitably depletes natural capital and if the depletion continues, natural capital will eventually be reduced to zero. The further beyond sustainable scale this depletion occurs, the more vulnerable are the multitude of ecosystem services provided by this fund of natural capital. And the longer the level of material throughput exceeds sustainable scale, likewise does vulnerability increase. Exceeding sustainable scale therefore has two major effects:

- the quantity and quality of ecosystem services begin to decline.
- the loss of ecosystem resilience gets worse as the duration and magnitude of exceeding scale continues.

*Implications of Exceeding Sustainable Scale*

Ecosystems are generally designed to be sustainable under a broad range of conditions and can continue providing some level of services even beyond sustainable scale. But if pushed beyond this range by human activities, they react in unpredictable ways. For example, changes are not necessarily linear with respect to the levels of stress applied. A doubling of atmospheric greenhouse gas concentrations beyond preindustrial levels will have more than twice the impact of a 50% increase. At some level of continued disruption, an ecosystem will flip into a different equilibrium state, and may do so quite rapidly. The ecosystem's natural resilience will assist in maintaining the old equilibrium, but as resilience declines, the likelihood of an equilibrium flip increases. Our knowledge of global ecosystems is totally inadequate to allow us to predict when such flips might occur. Therefore, the longer we exceed sustainable scale, the greater is the risk of reaching and exceeding maximum scale – the point of no return.

*Long Timelines*

The biophysical mechanisms that determine ecosystem variability operate over timelines that are very long by human standards. Both greenhouse gases and ozone depleting compounds continue to be active for up to a century after they are emitted into the atmosphere. The greenhouse gases emitted a century ago are influencing current climate patterns. Those emitted today will continue influencing climate stability for another 50 years or more. Anthropogenic ozone depleting compounds began acting almost immediately to upset the natural balance of atmospheric ozone creation and depletion, and those emitted today will continue their impact for a century or more. What we do over the next few years will have a long term impact on global ecosystems. The sooner we act to preserve and restore natural capital, rather than depleting it, the more we will enjoy the benefits of the associated ecosystem services, and the less we will have the time pressures of dwindling services and impending catastrophe. We are both richer and safer if we remain within the range of sustainable scale.

### Short Timelines

Ecosystems generally operate on timelines beyond the range of direct human experience. But they can change from one equilibrium state to another very rapidly. As material throughput increases exponentially, reducing doubling times, we may be misled to believe we have considerable time to remedy our ecological errors. We are relatively ignorant regarding the point at which the next doubling will push an ecosystem beyond sustainable scale. Not knowing just how short such a timeline might be suggests extreme caution in avoiding unsustainable scale.

## 7. Counterintuitive Solutions

### Perceived Solution is the Problem

Economic growth is considered by mainstream economists and most political parties as the solution to a variety of global problems. Economic growth is seen as the solution to poverty in both developed and developing countries, world hunger, over population, environmental degradation and a range of social problems. This belief is pervasive, crossing ethnic, religious and political boundaries. As one of the most significant ideas embedded in our global culture, it is difficult for believers in the benefits of economic growth to even consider that with all its apparent benefits, the downside of growth has the potential for devastation and irreparable harm.

*Limits to the Contributions of Economic Growth*

The concept of sustainable scale clarifies that continued expansion of the global economy can be "uneconomic" in the sense that costs may exceed benefits. This may occur even when traditional market signals obscure this net loss, by ignoring *externalized costs*. The concept of sustainable scale is compatible with the notion that economic activity is essential to human well being, but clarifies that there are limits to the contributions economic growth can make. The biophysical limits of ecosystem functioning which determine sustainability define those limits beyond which continued economic growth becomes destructive. By connecting the issues of economic growth and ecosystem sustainability, the concept of sustainable scale provides an important and distinctive perspective in sorting out these vital challenges.

# CHAPTER III
## Causes of Sustainable Scale Problems

The topic of scale is one of the themes that unifies different disciplinary perspectives. Phenomena of interest such as processes, patterns, individuals, and networks exist within a context that may vary in its dimensions, e.g. size, speed, complexity or other attributes. Studies of particular phenomena usually focus on events that occur within a particular combination of dimensions that defines a single scale at which empirical observations are made. Two considerations affect the choice of scale. On the one hand, there is the empirical reality of the phenomenon of interest, which may range across many scales. On the other, there is the subjectivity of our observations, which are by necessity tied to the scale or range of scales at which we can collect information. The objectives of scaling studies are to consider how our perceptions of phenomena of interest change as the scale of analysis changes and to try to assess objectively the multiscale nature of the phenomenon. When we consider the interactions of two systems, particularly those in which a cause-and-effect relationship exists, we are faced with the problem of understanding how scale influences the number and nature of those interactions. Social and ecological systems interact in many ways[55].

Here we consider the relevance of sustainable scale in the management of natural resources and of all the environmental problems we are encountering. These are growing as a consequence of a mismatch between *"what we want"* and *"what the ecological processes or natural resources are able to offer."*

---

[55] https://www.ecologyandsociety.org/vol11/iss1/art14/

Now, we address Sustainable Scale problems which are caused by:

*Excess Throughput:* biophysical limits of ecosystems have been exceeded by the material throughput in the global economy.

*Driven by Economic Growth:* economic growth inevitably involves material throughput; global trade increases throughput.

*Supported by Obsolete Public Policies:* growth oriented policies either support ecosystem degradation or fail to conserve critical ecosystem functions.

*Informed by Defunct Economic Theories:* the most policy relevant conceptual framework, modern economics, also ignores the impact of economic activities on ecosystem functioning.

*Encouraged by Dominant Myths:* Cultural myths about progress, science and human's relationship with nature all conspire to make ecosystem degradation acceptable.

*And Managed by Vested Interests:* Individuals and institutions in a position to make decisions about sustainable scale issues generally have a vested interest in the status quo.

## 1. Excess Throughput

*Excess Material Throughput*

Economic activities inevitably involve some level of material throughput. The important question is whether the level of throughput is sustainable with respect to *ecosystem functioning*. The evidence is clear that the volume and toxicity of material throughput is now in excess of what some ecosystems can bear and fast approaching the limits for other ecosystems.

*Material Throughput Breeds on Itself*

Not only does economic growth increase material throughput, but the goods and services produced inevitably lead to the use of yet more energy and materials. This occurs in two ways - goods and services are used, and eventually disposed of. Use involves additional material throughput. The global fleet of cars, trucks, buses, ships and airplanes become vehicles not only for transportation, but also for the throughput of fossil fuels and volumes of additional resources. Once disposed of as waste, the goods produced are often transported over large distances before being buried, incinerated or recycled, all of which requires more material throughput.

*Increased Quantities of Throughput[56]*

*The economic drivers:*

- in the coming decades, growing populations with higher incomes will drive a strong increase in global demand for goods and services.
- global gross domestic product (GDP) is projected to quadruple between 2011 and 2060, according to the central baseline scenario projected with the OECD ENV-Linkages model. By 2060, global average per capita income is projected to reach the current OECD level (around USD 40 000).
- production and consumption are shifting towards emerging and developing economies, which on average have higher materials intensity.
- the growing share of services in the economy will reduce the growth in materials use as the sector is less materials intensive than agriculture or industry.
- technological developments will help decouple growth in production levels from the material inputs to production.

*... of materials use:*

- global materials use is projected to more than double from 79 Gt in 2011 to 167 Gt in 2060. Non-metallic minerals, such as sand,

[56] https://www.oecd.org/environment/waste/highlights-global-material-resources-outlook-to-2060.pdf

gravel and limestone, represent more than half of total materials use.

- the materials intensity of the global economy is projected to decline more rapidly than in recent decades - at a rate of 1.3% per year on average - reflecting a relative decoupling: global materials use increases, but not as fast as GDP.
- recycling is projected to become more competitive compared to the extraction of primary materials.
- the strong increase in demand for materials implies that both primary and secondary materials use increase at roughly the same speed.

*...and its environmental consequences*

- more than half of all greenhouse gas (GHG) emissions are related to materials management activities. GHG emissions related to materials management will rise to approximately 50 Gt $CO_2$ - equivalents by 2060.
- fossil fuel use and the production of iron & steel and construction materials lead to large energy-related emissions of greenhouse gases and air pollutants.
- metals extraction and use have a wide range of polluting consequences, including toxic effects on humans and ecosystems.
- the extraction and use of primary (raw) materials is much more polluting than secondary (recycled) materials..

*Hidden Throughput*

In addition to the materials that are counted as part of economic activities, a considerable amount of throughput occurs which has no market value. This ecological rucksack, or hidden throughput, is generally many times the volume or weight of the target material. To produce a 10 gram gold ring, for example, requires the movement of some 3 tonnes of material. The rucksack for coal is at least five times greater than the weight of the coal mined. As the concentrations of valuable ores and minerals decline through depletion, the hidden throughput required to produce the same quantity of ore increases. While these hidden throughputs have no commercial value, they have major impacts on ecosystem functioning.

*Exponential Increases in Throughput*

The increase in most materials involved in economic activities approximate exponential growth. This means that the increase is proportional to what is already there. This is a much higher rate of growth than a linear increase. When exponential growth occurs, the quantities involved can double very quickly, depending on the rate. Even relatively low growth rates can therefore lead to fairly quick doubling. The longer such exponential growth continues, the shorter the doubling time. The implications for ecosystems is that exponential increases in material throughput will be at a level only half that required to push them beyond sustainable scale just prior to the doubling which produces this level. With substances involving exponential growth, a large safety margin is required to ensure optimal scale.

*Increasing Toxicity*

The EC inventory published by ECHA[57] is a copy as received from the JRC in 2008 on the founding of ECHA. It is comprised of the following lists:

- EINECS (European Inventory of Existing Commercial chemical Substances) as published in O.J. C 146A, 15.6.1990. EINECS is an inventory of substances that were deemed to be on the European Community market between 1 January 1971 and 18 September 1981. EINECS was drawn up by the European Commission in the application of Article 13 of Directive 67/548/EEC,

---

[57] https://echa.europa.eu/it/information-on-chemicals/ec-inventory

as amended by Directive 79/831/EEC, and in accordance with the detailed provisions of Commission Decision 81/437/EEC. Substances listed in EINECS are considered phase-in substances under the REACH Regulation.

- ELINCS (European LIst of Notified Chemical Substances) in support of Directive 92/32/EEC, the 7th amendment to Directive 67/548/EEC. ELINCS lists those substances which were notified under Directive 67/548/EEC, the Dangerous Substances Directive Notification of New Substances (NONS) that became commercially available after 18 September 1981.
- NLP (No-Longer Polymers). The definition of polymers was changed in April 1992 by Council Directive 92/32/EEC amending Directive 67/548/EEC, with the result that substances previously considered to be polymers were no longer excluded from regulation. Thus the No-longer Polymers (NLP) list was drawn up, consisting of such substances that were commercially available between 18 September 1981 and 31 October 1993.

The Database contains 106,213 unique substances/entries.

The consequences of these toxic substances in our air, water and land are becoming increasingly evident:[58]

- just over one third (35%) of *ischemic heart disease*, the leading cause of deaths and disability worldwide, and about 42% of *stroke*, the second largest contributor to global mortality, could be prevented by reducing or removing exposure to chemicals such as from ambient air pollution, household air pollution, second-hand smoke and lead.
- chemicals such as heavy metals, pesticides, solvents, paints, detergents, kerosene, carbon monoxide and drugs lead to *unintentional poisonings* at home and in the workplace. Unintentional poiso-

---

[58] https://apps.who.int/iris/rest/bitstreams/916484/retrieve

nings are estimated to cause 193,000 deaths annually with the major part being from preventable chemical exposures.

- the list of chemicals classified as human carcinogens with sufficient or limited evidence is long. Occupational carcinogens are estimated to cause between 2% and 8% of all *cancers*. For the general population it is estimated that 14% of lung cancers are attributable to ambient air pollution, 17% to household air pollution, 2% to second-hand smoke and 7% to occupational carcinogens.

- exposure to certain chemicals, such as lead, is associated with reduced neurodevelopment in children and increases the risk for attention deficit disorders and intellectual disability. Parkinson's disease has been associated with exposure to pesticides. Other links between *mental, behavioral and neurological disorders* are suspected; evidence, however, is more limited.

- air pollution and second-hand smoke are risk factors for *adverse pregnancy outcomes* like low birth weight, prematurity and stillbirths. Antenatal exposure to second-hand smoke for example was estimated to increase the overall risk for stillbirths by 23% and for congenital malformations by 13%. There are, furthermore, potential links between various chemicals and adverse pregnancy outcomes or congenital malformations, though evidence is limited.

- *cataracts*, the most important cause of blindness worldwide, can develop from exposure to household air pollution. Exposure to cookstove smoke was estimated to be responsible for 35% of cataract disease burden in women and 24% of the overall cataract disease burden.

- second-hand smoke and air pollution are also responsible for 35% of *acute lower respiratory infections*, including pneumonia, bronchitis and bronchiolitis, the most important cause of mortality in children, and are also linked to upper respiratory infections and otitis media.

- more than a third (35%) of overall *chronic obstructive pulmonary disease* (COPD) burden is caused by exposure to chemicals in second-hand smoke, air pollution or occupational gases, fumes and dusts. Second-hand smoke and air pollution can induce reduced lung function and a predisposition for pulmonary disease in unborn and young children.

- second-hand smoke and air pollution can lead to the development of, and increased morbidity from, *asthma*. Air pollution additionally provokes asthma exacerbations and increases related hospital admissions. Asthma from occupational asthma genes is among the most frequent diseases related to the workplace.

- over 800,000 individuals die from *suicides* every year.[59] About 20% of suicides could be prevented through restricting access to poisons (estimate based on expert survey and limited epidemiological data). Self-poisoning with pesticides is the main means of suicide in India, China and some central American countries.

*Chemicals and air pollution*

Air pollutants from ambient and household sources are a mixture of many components including, for example, carbon monoxide (CO), sulfur dioxide ($SO^2$), nitrogen oxides (NOx) and particulate matter, the last containing substances such as acids, organic chemicals, metals, soil and dust particles. The way chemicals are managed can directly contribute to air pollution. One example is the use of pesticides in agriculture, which can volatilize and suspend into the air when sprayed. Phasing out leaded

---

[59] https://www.who.int/mental_health/prevention/suicide/suicideprevent/en/

gasoline has reduced the amount of airborne lead. However, the largest sources of air pollution are combustion and other processes from energy generation, industry and transport. Nevertheless, because of the chemical composition of air pollution which can vary to a large extent depending on prevailing pollution sources, the assessment of health hazards from these chemicals remains important. The scale of hazardous substances used in economic activities is threatening local, regional and global ecosystems. To operate within sustainable scale the rate of emissions for such substances would have to be no greater than ecosystems can recycle, absorb, or breakdown to harmless components. Unfortunately, just the opposite is happening. For example, under normal levels of acidification soils are able to absorb heavy metals, binding and sequestering them safely. At higher levels of soil acidification (as caused by acid rain), soils loose this capacity. This allows the increasing emissions of heavy metals to enter our water sources, and eventually our food supply.

*Wasteful Processes*

The flow of material through the various stages of economic activity are extremely wasteful. A study[60] prepared for the Australian Department of the Environment and Energy reports that the country in 2016-17 generated an estimated 67 million tonnes (Mt) of waste, including 17.1 Mt of masonry materials, 14.2 Mt of organics, 12.3 Mt of ash, 6.3 Mt of hazardous waste (mainly contaminated soil), 5.6 Mt of paper and cardboard and 5.5 Mt of metals. This is equivalent to 2.7 tonnes (t) per capita. There was about 54 Mt of 'core waste' – that managed within the waste and resource recovery sector (2.2 t per capita). This comprised 13.8 Mt (560 kg

---

[60] https://www.environment.gov.au/system/files/resources/7381c1de-31d0-429b-912c-91a6dbc83a-f7/files/national-waste-report-2018.pdf

per capita) of municipal solid waste (MSW) from households and local government activities, 20.4 Mt from the *commercial and industrial* (C&I) sector and 20.4 Mt from the *construction and demolition* (C&D) sector. Over the 11-year period for which data is available, waste generation increased by 3.9 Mt (6%). Assessed on a per capita basis, waste declined by 10% over this timeframe. MSW generation fell by 10% per capita and C&I waste by 8% per capita, while C&D waste grew by 2% per capita.

Thanks to the Government of Australia. Many other countries are far from publishing their data.

## 2. Economic Growth

*Economic Growth Drives Material Throughput*

Economic growth inevitably involves material throughput. It is this intimate connection between our current model of economic growth and ecosystem destruction that is at the core of sustainable scale problems. Economic growth involves selling ever more products and services, to ever more people, each of whom consumes ever more of these goods and services. Each of these major drivers of consumption – population, technology and per capita consumption - is viewed by decision makers as a positive and desirable contributor to economic growth. It is these very same drivers that disrupt life supporting ecosystems and threaten ecosystem collapse. The many attractive and beneficial consequences of economic growth make it difficult to acknowledge and accept that there are also significant harms inextricably linked to the growth process.

*The Downside Of Economic Growth*

High levels of public concern for the environment are an indication of a broad recognition connecting economic activities and a deteriorating environment. There is little public or political recognition, however, of the potential magnitude of these negative impacts, our current status with respect to exceeding sustainable scale boundaries, and the related consequences. Acknowledging this intimate connection between economic growth and the potentially dire consequences of exceeding sustainable scale is essential to dealing with this critical challenge.

### The Psychology of Scale

Making the connection between economic growth and unsustainable scale is a shock to the psyche. It is a bit like learning that someone close to us, a spouse, a best friend, or a business partner, is actually doing us serious harm. A common reaction is denial. But if the evidence of the harm is reasonable the problem is too powerful to ignore for long. Investigating the situation reveals the harm, while real, is not intentional, and can be avoided.

### Support for Economic Growth Broadly Institutionalized

Economic growth occurs within a context influenced by public policies, economic theories and prevalent social ideas and values. How any harm done by economic growth is viewed, is also influenced and shaped by these same powerful forces. For varied reasons, each of these major influencers of economic growth is oriented toward promoting even more growth and ignoring or minimizing what could be potentially catastrophic effects.

### Links Between Growth and Justice

A striking characteristic of modern economic growth is the enormous disparities in the distribution of benefits it provides. This is a justice issue of global proportions. Sustainable scale challenges cannot be resolved in isolation from this issue of distribution, as mainstream policies for redistribution of wealth presuppose an economic product that is growing forever.

## 3. Obsolete Public Policies

*Economic Policies Set Growth as the Priority*

It would be difficult to find any governmental body, at any level, in any nation, which did not have economic growth as a key if not central goal of its administration. Economic considerations are generally given the highest priority in public policy decision making. If there are conflicts with social or environmental concerns, political decision making generally favors economic interests. Higher policy priority will have to be given to maintaining ecologically sustainable scale to avoid the inevitable consequences of exceeding scale. The overarching priority of economic growth was easily the most important idea of the twentieth century.[61]

*Money Policy Drives Economic Growth*

One of the most fundamental policies of any government has to do with how money is created. This is known as seignorage, acknowledging that such a task was originally the sole right of the monarch. In most developed countries this right is now in the private sector and is accomplished by requiring economic growth for new money to be created. Money creation, economic growth and a rapid advance on sustainable scale boundaries are intimately connected in reality if not in design, by public policies around the world. Scale problems will not be solved without adjusting money policies.

*Business Policy Encourages Growth*

The term "open for business" has become a cliche for cities, regions and nations around the world. Governments encourage business development through a variety of laws and regulations, and by generally facilitating business expansion regardless of the environmental and social consequences. In many jurisdictions, corporations are required by law to ignore these consequences in the interests of

---

[61] McNeill, J. R. "Something New Under the Sun". New York: W. W. Norton Co., 2000: p.337.

shareholder return. These *externalized costs* obscure an accurate accounting of business operations. Business practices which unnecessarily contribute to material throughput are encouraged and supported by numerous government policies and programs. They are incompatible with resolution of the scale problem. Alternative business approaches are available.

*Trade Policy the Vehicle for Economic Growth*

Most nations seek economic growth and prosperity through trade with other nations. Most international trade agreements make economic growth priorities clear; social or environmental issues cannot legally interfere with the opportunity for economic growth. These trade agreements, by their nature, encourage further increases in material throughput. Policies which fail to consider the impact of increased international trade on the sustainability of ecosystems pose a serious threat.

*Environment Policy Off the Mark*

The vast majority of environmental regulations by governments and voluntary environmental measures by businesses and consumers, occur without identifying sustainable ecosystem impacts (the same is often true for environmental organizations). If clear targets are set at all, they are often the result of political negotiations rather than based on principles of sustainability. Such an approach places greater emphasis on the palpable unpleasantness of environmental problems, rather than on their less visible threats to ecosystem sustainability. Environmental policies that neglect to identify the capacities of ecosystems to continue functioning under ever increasing levels of material throughput are of limited usefulness.

*Growth as the Solution to Policy Challenges*

Economic growth has become the dominant policy instrument partly because it is believed to hold out promise to solve a variety of social, environmental and economic problems. Economic theory creates great faith in the market's ability to generate well being. Economists advocate letting the market do its work as the most efficient approach to benefits for all. These beliefs are firmly held, frequently repeated and energetically defended, despite considerable evidence to the contrary. These economic theories supporting the growth priority have serious flaws, drawing into question both their validity and their role in policy formulation. We will see alternatives available in Volume III.

*Reduced Size and Scope of Government*

The trend to reduce the size of governments in the world's major nations means that the mechanisms and procedures for developing sound public policies are also reduced. Here, the potential loss of democratic representation must be taken into account. What remains is largely focused on promoting economic growth as the goal which will allow all other major problems to be resolved. There are scarce public resources available, and even scarcer political will, to consider the harmful impacts of this growth. There is even less public policy space available to grapple with the scale issue, even though it touches directly on each of the main areas of government responsibility. National governments have the responsibility to address scale issues and need to commit sufficient resources to do so.

*Government Responsibilities*

The ideal for most modern democracies is to ensure that the basic needs of all citizens are met and that all are able to reach their potential. Governments' responsibilities in areas of education, healthcare, environmental protection and social justice, are affected by their ability to tax their citizens. Taxes rise with increases in economic growth, so governments see economic growth as the means of fulfilling their responsibilities. However, when growth becomes the overriding goal, policies are implemented which support growth without fulfilling these other obligations. Economic growth considered as *a means* to citizen well being (rather than as *an end* in itself) would be a policy change compatible with sustainable scale. The importance of sustainable scale for well being should be reflected in public policies.

## 4. Defunct Economic Theories

*Economics for an Empty World*

Economic theory has come to be the major discipline which influences policy makers at all levels of government. Today's economic theory has its roots in the 19th century when the world was relatively empty in terms of people, throughput and wastes. The theory has changed over the course of the last century, but unfortunately, has not adjusted to the fact that the world has become fuller of people, material things, wastes and toxicity. Today, several key theoretical concepts from neoclassical economics are unsupportable, yet they continue to influence government decision makers. Several of these very dubious concepts, and the absence of others are particularly relevant to scale.

*The Power of Market Mechanisms*

The market economy is the playing field of neoclassical economics. What the market does best, is most efficient at, is the allocation of resources based on supply and demand. Supply and demand are calculated to determine the point at which the production of a certain good or service no longer reaps a profit. Microeconomics is primarily about supply and demand. It determines when allocation of labour, capital and resources is no longer efficient because the supply is outstripping the demand. This very clear "when to stop rule" maintains an efficiency in the market by avoiding wasteful allocations. The "when to stop" rule influences the amount of material throughput on the microeconomic level – but the regulator is profit, not ecosystem integrity or social justice.

*The Missing "When to Stop Rule"*

There is no theoretical underpinning for a *"when to stop rule"* for aggregate macroeconomic activity in neoclassical economics. Lacking such a conceptual framework, neoclassical economics is not able to indicate when the global allocation of labour, capital and resources should stop because it is *"uneconomic"*, that is, when costs exceed benefits. There is no method for measuring the point at which this uneconomic activity begins; nor is there even a concept that suggests this point exists. However there is considerable evidence from a variety of scientific disciplines that indicate this *"when to stop"* question should be asked if global ecosystems are to survive.

*Limits of the Market Economy*

Despite its many strengths, neoclassical economic theory requires more of the market than it can deliver. Some of these limitations may be fixable; other problems require reformulation of theory and development of new techniques. Some of the more serious market failures include:

- *Monitoring Well Being:* the policy emphasis on economic growth has led to increases in Gross Domestic Product (GDP) being regarded as increases in human well being. But GDP is a measure of gross production and consumption and does not assess the value of either to human well being. Many aspects of production and consumption contribute to human misery and ecological degradation, but nonetheless add to the GDP. Alternative measures being developed attempt to clarify what is a "good" or a "bad" type of production or consumption (e.g. the ISEW or Index of Sustainable Economic Welfare; and the GPI, or Genuine Progress Indicator). Comparisons of these alternatives to GDP over time show that GDP continues to grow while these other measures either decline or remain stable.

- *Suppressing Externalities:* in economic theory an externality is an unintended cost or benefit to a party not directly involved in a particular market exchange. Ecological degradation is considered an externality, such costs being borne not by the market participants, but by the general public, future generations and non-human species. If those involved in the relevant market transaction bore these externalized costs, then many currently profitable transactions would become uneconomic and cease. The policy emphasis on economic growth has led governments to allow or even encourage businesses to externalize costs as a way of facilitating continued growth. These practices serve to increase material throughput and create not only serious local and regional environmental problems, but also serve to alter global ecosystem services upon which we depend.

- *Non-Market Goods and Services:* many goods and services that are either essential for survival, or contribute greatly to human well being, cannot become part of the market system because they are non-rival. This means they can be enjoyed without interfering with others' enjoyment of them. Some of these non-market goods and services are provided by ecosystems (eg. UV radiation protection, or scenic beauty). Other non-market goods or services are provided by social systems (eg. either as public goods like democracy; or private services like friendship). Economic theory encourages practices which attempt to price these goods and services so that they can enter the market and contribute to growth. But doing so ignores the issues of sustainability and social and ethical values involved with these non-market goods and services. Price does not, and cannot, capture everything of value. Attempting to bring non-market goods and services into the market economy only serves to increase material throughput without adding to human well being.

- *Maximizing the Wrong Thing:* Economic theory focuses on maximizing utility as measured by consumption. Maximizing consumption, it is believed, is the way to satisfy the broadest range of insatiable human wants. But it is the satisfaction of needs, not consumption itself, which is the appropriate goal. And humans needs are much more complex than those which can be satisfied in a market exchange. By focusing on consumption of market-based goods and services, neoclassical economic theory encourages material throughput and provides false hope of meeting the full range of human needs with such consumption. Combined with a growing global population and increasing technology-driven throughput, neoclassical economic's misplaced maximization of consumption is challenging ecosystem integrity.

*Continuous Growth*

Neoclassical economics is based on generating profit to reinvest in more production, or research and development as an intermediate step to the same end. All profit is ultimately based on a surplus from agriculture and extractive industries – mining, logging, ranching and fishing. The money supply expands as profit based on this surplus grows. Any other expansion of the money supply is inflationary. Even speculative investments are ultimately based on real resources. If speculation is overly optimistic, inflation occurs, and any apparent increases in the money supply are illusory. Either an "adjustment for inflation" is required, or a more painful "market correction" occurs. Another way of thinking about the importance of extractive industries is to consider that no profit is possible without material throughput.

*Ignoring Nature*

Despite this obvious and fundamental connection of economic growth to the natural world, neoclassical economic theory ignores the role of nature in the economic production process. When first developed, economic theory spoke about the factors of production in terms of land, labour and capital. Modern economic textbooks focus only on labour and capital. *"Land"* (meaning all natural resources) is regarded as unessential because capital is considered to be an adequate substitute. This elimination of land and the elevation of capital in economic theory, came about as economists noted that substitutes could be found for one resource which became scarce by investing profit in another resource (coal as a substitute for wood as fuel; later petroleum became a substitute for coal).

These transitions gave rise to the illusion that with enough money, anything is possible.

*Pillars of Economic Growth Theory*

Technological progress, along with substitutability, is the other foundation pillar of continuous growth theory. It is assumed that with enough money, research will always provide a technical solution to either increase efficiency, or to substitute for some valuable resource which becomes scarce through overuse.

*Technological Progress as a Perpetual Motion Machine*

Neoclassical economists believe that technological progress will allow for the every increasing production of goods and services. More goods and services mean more profit, which can then be reinvested in more production. But in addition to conferring many benefits, technological change has also led to ever increasing amounts of material throughput. This occurs even when increasingly efficient use of natural resources stems from technology. Increased efficiencies can lead to gross increases in throughput because the greater efficiencies allow us to use more of the resource. Coupled with the increase in human population and the spread of the market economy, gross increases in throughput are common despite more sophisticated technologies (eg. sidebar increased fuel efficiency leading to more cars and miles driven). Indeed, increases in some technologies allow others to produce ever increasing levels of throughput (eg. cheap energy leads to increased transportation, which leads to increased trade, which leads to more material throughput).

*Problems with the Pillars*

Substitutability is limited in the real world if not in neoclassical economic theory. Some resources, such as water have no substitutes, yet are essential to support life. But many of today's major environmental challenges are not matters of resource substitution, but of overwhelmed sink capacities. What substitutes are there for the atmospheric ozone layer, biodiversity, or the carbon cycle and greenhouse effect? The neoclassical economist's claim that capital is a substitute for natural resources ignores the inherent connection between nature and profit. The biodiversity and other ecosystem services provided by nature are the source of all our wealth.

*Violating the Laws of Physics*

Assuming that one piece of nature is substitutable for another (because it can be transformed by technology) ignores the second law of thermodynamics. Some pieces of nature are more valuable precisely because of their unique characteristics which cannot be duplicated by technology (coal combusts at a certain temperature and produces energy, a rock does not). It is these unique characteristics produced by nature that are important to us in terms of the services they provide. It is also these unique characteristics that are destroyed in the production process (once the piece of coal is burned, we can never again use the dissipated heat and ash to generate more energy). According to the second law of thermodynamics this loss of unique characteristics in the production process is inevitable, and always results in a degradation of the original resource. Technology may delay this process for a while, but inevitably the second law will apply. The bottom line is not profit, but the laws of physics which determine the limits to substitutability, technological progress, and ecosystem services.

*Continuous Economic Growth Incompatible with Sustaining Ecosystem Integrity*

There are no substitutes for the enormous variety of critical ecosystem services we rely on for our well being. Nature provides these services on its own, and very economically, as long as human generated material throughput does not interfere. This is what the process of evolution of species and ecosystems is all about. Not only are there no substitutes for these critical services, but their complex, dynamic and emergent properties means that no amount of technological progress will be able to reproduce even a fraction of what these natural systems provide, regardless of the amount of money, research and energy we devote to such tasks. If we continue to rely on the fallacious notion of substitutability, and the illusory promise of technological progress, the two pillars of continuous growth theory, we will ensure the speediest possible route to ecosystem catastrophe.

*Human Hubris*

Human creativity is one of the wonders of the universe. It is no simple matter to balance this powerful gift with a humility which reflects an appreciation of nature's dynamic complexities and limits. Our ability to achieve an appropriate balance will determine whether our economic activities can achieve a sustainable scale. The emergent and non-linear properties of complex ecosystems means there is an irreducible uncertainty inherent in our understanding of these systems. Striving to reduce our ignorance is an inherent and endearing human trait. Yet, acknowledging the biophysical limits of nature is essential for our survival. Our greatest challenge is to continue exploring these limits while still respecting their inevitable implications.

## 5. Dominant Myths

*Basic Beliefs*

Every civilization relies on commonly held assumptions about how the world works and what is of greatest value. Most of these common beliefs remain untested, but influence decisions by being accepted as basic assumptions. If beliefs are erroneous they can easily restrict a society's ability to solve problems.[62] Such distorted worldview influence the present global culture, and some of these central but erroneous beliefs have an impact on the scale issue.

*The Relationship between Humans and Nature: Dominion or Stewardship?*

The European perspective on the relationship between humans and nature has become the dominant global view. Based in Jewish-Christian traditions is the belief that humans were created to have dominion over nature and all its creatures. In the earliest chapters of Genesis there is reference to this human dominance, accompanied by a divine proclamation for humans to subdue nature. Nature is there to serve humans and be subordinate to human manipulation. This notion was reinforced throughout the history and expansion of European thought across the globe. It is a notion that prevails today in Western culture, as continuing attempts are made to subjugate nature and remove its messy intrusions upon the regimented and sanitized character of "modern" life.

---

[62] Diamond, Jared. Collapse: How Societies Choose to Fail or Succeed. New York: Viking, 2005.

*One with Nature*

Eastern and indigenous cultures developed a different view of the relationship between humans and nature – one of human responsibility to care for and preserve the natural world. Nature itself is viewed as sacred, and humans are viewed as part of and dependent on the natural world. By caring for nature, humans are caring for god's creation and themselves as well. Such a view has also been represented in most major religions, but relegated to a minority tradition.

*The Notion of Progress*

Today the notion of progress is prevalent in most societies around the world. This is a relatively new concept originating in 17th century in Europe with the emergence of modern science and technology. Previously, societies were viewed as either having no particular directions, or as decaying from a previous state of grace. The idea that things will inevitably improve because of increased knowledge and technologies is new, and reinforced by the rapid rate of scientific discoveries and technical innovations that are brought to market. The extent and rapidity of such changes are accepted as evidence that continued progress is inevitable.

*Faith in Science and Technology: Technology Will Save Us*

The incredible advances in science and technology over the last century are truly remarkable. This explosion of knowledge is unprecedented in human history and its application in the everyday lives of hundreds of millions of people has come to be taken for granted. The range and complexities of these discoveries reinforce the belief that humans can understand and subjugate nature and that whatever problems societies face, science and technology will provide solutions.

*Questioning the Dominant Myths*

Taken together with the prevalence of the market economy, these dominant cultural worldview present formidable obstacles to recognizing the scale issue as a potentially serious threat. The idea that humans have pride of place in the universe, with an obligation to exploit nature to glorify a deity, and have an unlimited intellectual capacity to create technologies as a means of doing so, along with the material evidence that economic growth has benefited hundreds of millions of people, provides powerful incentives to ignore the scale issue. These ideas support and reinforce each other, and have shaped many key institutions that characterize modern societies. There is no "when to stop" rule in either neoclassical economic theory, or the Book of Genesis.

*New Myths to Consider*

To think about scale issue seriously, consideration must be given to the following ideas, each of which runs counter to a now dominant myth:

- humans are part of nature and rely on nature for survival.
- humans have a special responsibility in protecting nature.
- current science is limited in its ability to understand the complexities of nature.
- technologies can create irrevocably destructive environmental and social problems.
- economic growth has the potential to irrevocably destroy parts of nature which are required to support life.
- societal progress is not a given; global civilization could collapse.

The history of the rise and fall of civilizations is largely a history of solving major social challenges by pushing a worldview that works to its logical limits, and failing to foresee those limits and how they might be avoided.[63] To ignore the scale issue is to fail to learn from this history and to commit the same error today.

*Scale and the Survival of Civilization*

In the past, civilizations would fail in one place, but new ones would emerge elsewhere. The global extent of modern civilization, connecting all the earth's major ecosystems through ever expanding economic activities, threatens to eliminate the prospects for such renewal. It is both the pervasiveness of the problem, and the irrevocable destruction that our economic activity is capable of, which makes the scale issue unprecedented and of such vital importance.

## 6. Vested Interests

Institutions and practices endure because certain individuals and groups benefit from them. Generally, those that benefit do what they can to ensure the institutions and practices continue and become stronger. There are many beneficiaries of the current paradigm of economic growth, each of whom actively supports it.

*Self Interest as a Core Virtue*

Personal self interest has been exalted to the status of a core virtue in Western societies. Derived from economic theory and reinforced by the Western notion of individualism, the pursuit of personal self interest has become a justification for consumption and greed. Self interest has also become a justification for ignoring, and even ridiculing, the pursuit of the common good. Economic theory claims that the pursuit of personal self interest will generate the greatest good for the greatest number. From this perspective not only is independent pursuit of the common good unnecessary, it is actually considered counterproductive as it may interfere with the market mechanism. It is this belief that underlies efforts to reduce

---

[63] Tainter, Joseph A. *The Collapse of Complex Societies*. Cambridge: Cambridge University Press, 1988.

the size and role of government, the key institution responsible for ensuring the common good.

*Businesses and Investors*

The creation of wealth in the last half century is unprecedented in human history. The wealth created is not evenly distributed but disproportionally allocated to those who successfully operate businesses. Investors who provide financial backing also reap disproportional rewards. Many employees at more senior levels and unionized workers also share in the benefits. Those who benefit most generally play a much more active role in maintaining the status quo. They have the disposable financial resources to spend in furthering their cause. Individual corporations, as well as business associations and lobbies, and business supported policy institutes, expend considerable resources in supporting and expanding the dominant paradigm of economic growth.

*Governments and Politicians*

Politicians and government officials the world over increasingly view economic growth as the key to electoral success. Factors reinforcing this phenomenon are the increasing reliance on election campaign contributions from the business sector, and the revolving door feature of appointments at senior levels of government and business.

*Professional Disciplines and Institutions*

Coevolving with economic theory were a variety of related disciplines such as accounting, finance, law, corporate management and advertising. These disciplines are intimately intertwined with economics and each other, creating a complex web of interconnections that are mutually reinforced on a daily basis. Each of these disciplines relies on the continued growth of economic activity, and thus has vested interests in such growth.

*The Consuming Elite*

Across the globe there is a social stratum that enjoys the major benefits of economic growth. The majority live in developed countries but substantial numbers live in undeveloped areas as well. They generally associate their happiness with economic growth and enjoy various luxury goods and services. Collectively, they represent approximately 10% of the world's population, but control quite the total of the world's wealth. They are either business leaders, politicians, or professionals, or have access to this upper echelon of society through the ballot or their role as consumers. Such people have a vested interested in continued economic growth. This interest is justified and reinforced by the neoclassical economics dictum that social benefits accrue to those who pursue their own self interest through individual market decisions.

*Comfortable Consumers*

In addition to the consuming elite, the median annual household income worldwide is $9,733, and the median per-capita household income is $2,920, according to new Gallup metrics. Vast differences between more economically developed countries and those with developing or transitional economies illustrate how dramatically spending power varies worldwide. Median per-capita incomes in the top 10 wealthiest populations are more than 50 times those in the 10 poorest populations, all of which are in sub-Saharan Africa[64].

*Majority Excluded*

The other 72% of humanity lives below the poverty level of Western Europe; approximately 3 billion subsist on $2 or less a day. The common wisdom is that the vast majority would like to enjoy the material benefits of the more well to do. If there is any justifica-

---

[64] https://news.gallup.com/poll/166211/worldwide-median-household-income-000.aspx

tion for continued economic growth it is to meet the basic needs of this group. Given their large numbers, it is essential to do so in an ecologically sustainable way.

# CHAPTER IV
## Dialogue with the Skeptics

Scepticism interest us because it is an approach that underlies much of what we think about knowing, and also because much of this book is about how computational tools can help us in thinking through. Sceptical attacks on knowing have provoked many of the most imaginative attempts to methodically ground human knowledge. Scepticism is a tradition of systematically questioning any certainty, and, as in the case of Descartes, our belief in any reality. The problem with scepticism, as is pointed out by Hume[65], is that when the conversation is over you end up not being able to believe in any knowledge, even whether doors or windows are the best exits from a building.

More important, the tradition of sceptical questioning as found in many philosophical works, creates space for agile interpretation instead of trying to permanently solve interpretive questions.

It is therefore time to ask us some questions to try to reflect on the issues, abandoning the choice of questioning and choosing, instead, that of solving.

### 1. About legislation

*Skeptic:* Won't environmental protection legislation ensure our security? The extent of such legislation seems to be growing, as well as the number of international environmental treaties. Surely, all this attention by legislators and scientists will ensure sustainable scale is not exceeded.

*Response*: It is highly unlikely that current environmental legislation will adequately deal with the sustainable scale issue. The que-

---

[65] Dialogues Concerning Natural Religion (1779)

stion of sustainable scale is not even being asked by governments. The appropriate ecosystem indicators are not being monitored, and there are no formal attempts to identify either the biophysical or ethical and social boundaries of optimal scale. Nor are any efforts being made to educate the public so that they can participate in the development of a consensus regarding optimal scale. Most environmental legislation deals with *"end of pipe"* solutions, focusing on minimizing the impact of wastes rather than preventing the waste in the first place. Such legislation is confined to dealing with well specified environmental problems, rather than complex ecosystem effects. While science is learning a great deal about ecosystems, there are few instances where the specific measures are clear enough to allow legislation. Prevention is generally more desirable and less costly than clean up, especially as many of the ecosystem impacts may be unknown, or difficult or costly to detect and monitor. Experience with atmospheric ozone and greenhouse gas emissions indicates that scale problems can arise fairly quickly. Exceeding sustainable scale in terms of disrupting the underlying mechanisms of any single major global ecosystem (rather than merely altering it a bit), could cause a chain reaction in other major global ecosystems. This would make remediation of any kind virtually impossible. The current magnitude and potentially irrevocable nature of scale problems require urgent attention, but they are being ignored at policy levels. Environmental legislation is also subject to political whim. Various regulations have been rescinded by unsympathetic legislators when new parties come to power. In addition, even the best environmental legislation can be subverted by weak implementation or enforcement. There are many factors pushing legislators to ensure economic growth is a policy priority and environmental legislation of any kind is often viewed as an obstacle to such growth. Many scientists and scientific organizations have, in fact, spoken out clearly about the potential dangers of many current practices and policies. These have largely been ignored or dealt with in a token fashion by legislators. With respect to *Climate Change*, for example, the international treaty signed by many countries has an emissions reduction target that will result in concentrations of atmospheric carbon that scientists have warned will have dramatic consequences. The recommendations made by scientists regarding much higher emissions reduction targets were overlooked in favor of a politically acceptable target. Clearly, legislation is required to deal with concerns regarding sustainable scale. But significant changes in the development of public policies are required to adequately address the issue.

## 2. About take action

*Skeptic*: If the science regarding the potential dangers of exceeding scale scale are so clear, wouldn't legislators be forced to pay attention? There must be considerable uncertainty about the "danger signals," for such warnings to be ignored.

*Response*: Scientists have been very clear about the potential dangers of current practices, even if they have not always expressed their concerns in a sustainable scale framework. The specific scientific evidence may not yet be definitive, but some of the most stable and powerful scientific laws clearly demonstrate the impossibility of continuous economic growth. Legislators generally attempt to balance a broad variety of perspectives in their policy deliberations. If they are lobbied from conflicting sources, they tend to give priority to economic interests more than any other. Scale issues are of such dire importance that this approach could be disastrous. The tactic may work for parties with conflicting vested interests, but the basic laws of science cannot be suspended or mollified by political compromises. Scientists may also disagree on some of the data and its implications. The highly technical nature of much of the scientific data makes it difficult for legislators to determine which position is valid, even when there is an overwhelming consensus among the majority of scientists. If there are any dissenting voices among scientists, legislators often simply avoid a difficult decision until a definitive conclusion is reached. However, science rarely produces unanimity, especially in an area as novel and complex as ecosystem functioning. Scientists also tend to be very conservative in their statements so as not to overstate the facts. Legislators may interpret this approach as reflecting more uncertainty than is warranted. Sustainable scale problems of a global nature are unprecedented from both scientific and policy perspectives and miscommunication between these sectors is almost inevitable. Uncertainty and debate are normal aspects of science and should not be a justification for avoiding an issue as potentially catastrophic as exceeding scale. The uncertainty, after all, goes both ways. There is significant if not conclusive scientific evidence that some sustainable scale boundaries have already been breached, and that others are imminent. The uncertainty which exists should serve as a motivator, not an inhibitor, for ensuring sustainable scale is not exceeded with all its disastrous consequences. The unprecedented nature of the policy decisions that sustainable scale issues require means that legislators are on the very lowest levels of a steep learning curve. The danger of waiting for definitive scientific answers, and/or unambiguous danger signals, runs the risk of exceeding sustainable scale before adequate attention is given to preventing such overshoot. The one opportunity for prevention will be lost; there will be no opportunity for remediation. There is sufficient scientific evidence and consensus now available to develop preventive policies that could deal with scale issues in a constructive and economical manner. In addition, those areas where sustainable scale has already been exceeded need urgent attention even those the current impacts may be small or uncertain.

## 3. About the cost

*Skeptic*: Wouldn't implementing such policies now be costly and premature, especially if we are not really sure exceeding scale has already occurred, or is imminent?

*Response*: The basic question is what kind of error is tolerable. If policies are implemented that address potential sustainable scale problems, and we later discover they were premature, then at most we will have made unnecessary expenditures. However, if we fail to make the necessary changes in policies, regardless of costs, we run the risk of global catastrophe.

One of the unique aspects of this issue is that both the stakes and the uncertainty are so high. Exceeding sustainable scale in even one major global system could trigger major ecosystem degradations in a short period of time. Exceeding sustainable scale in one global system makes it more difficult to both understand and improve problems with other major systems. The scientists who study these complex global systems are in the best position to understand the potential dangers we face, and there is considerable consensus on many of these issues to warrant action earlier rather than later.

Furthermore, while the adjustments needed to the global economy are significant in order to avoid exceeding sustainable scale, there could be considerable collateral benefits to such policies in social, environmental and even economic terms. A thorough comparison of these pros and cons are not being examined at the present time. Yet this is precisely the kind of public debate that is needed globally because of the seriousness and irrevocable nature of sustainable scale problems.

## 4. About interpretation

*Skeptic*: Isn't the sustainable scale issue just another form of the "Limits to Growth" arguments made by the Club of Rome decades ago? These arguments have been proven wrong, so sustainable scale concerns really have no basis.

*Response*: It is true that some of aspects of the Club of Rome's early work was inaccurate and had some weaknesses. However, the basic thesis that infinite growth in a finite world is a physical impossibility was and is still valid. The rapid rate of economic growth and ecosystem disruption is continuing and unsustainable. Many of the weaknesses of the earlier work have been corrected in another publication, *"Limits to Growth: the 30 Year Update."* In fact, the basic conclusions of the original work have been well supported by subsequent events. Using the improved computer modeling in the latest report demonstrates that only by substantially reducing the assumptions concerning economic growth can a sustainable future scenario be realistically anticipated.

## 5. About economy

*Skeptic*: Solving environmental and other global problems like poverty and hunger require funding. We need economic growth to generate the wealth to solve these problems. It is the more successful economies of the world that have the cleanest environments. If we stop growth we are effectively giving up trying to solve these problems, and that would be unfair to the disadvantaged of the world.

*Response*: There are several points to make in responding to these views:

- all wealth is not to be measured in financial terms. Ecosystems are the source of all our wealth and make significant contributions to our well being. Simply because most ecosystem services have no market value does not mean they are worthless. Part of the problem with current public policy is that it focuses almost exclusively on financial factors and largely ignores these other values. Ecosystem health is the basis of a sound economy. While both are important, protecting ecosystem health is a greater priority than protecting economic growth. The value of healthy ecosystem functions is too poorly understood by decision makers for them to be concerned about threats to these unique and finite life-support services.

- funding is definitely required for environmental protection. The cost of cleaning up the toxic wastes sites in North America alone is estimated to be in the hundreds of billions of dollars. But does it make sense to generate the wealth required to undertake this cleanup from economic activities that generate yet more such toxic wastes? It is the kind and amount of economic growth that requires careful review and redirection that can assist in dealing with the sustainable scale challenge.

- it is a fallacy to attribute cleaner environments in successful economies with economic growth. First of all, many environmental improvements do not require huge expenditures. The environmental legislation that has contributed to the cleanup in developed nations has often resulted in reduced operating costs for business and industry. In addition, a significant portion of these improved conditions have at least as much to do with the export of dirty industrial work to less developed countries, where environmental protection is weak or nonexistent. Failure to recognize that ecosystem threats are global means that these tactics are at best a temporary reprieve. It is the sheer volume of material throughput, and the toxic nature of a significant portion of it, that is the problem – not the country of origin.

- dealing with the sustainable scale issue does not require the end of all economic growth, but it does require change. Economic activity based on sustainable business development principles such as zero waste, and production processes that only make goods that are recyclable or compostable, and that only use recycled materials, could go a long way to remaining within sustainable scale. These principles are designed to increase resource productivity and decrease material throughput, and will be important components of operating an economy within sustainable scale.

A sustainable economy is an ideal to strive for. Technologies are currently available to make radical improvements, of 400% or more[66], in resource productivity and waste reduction. But supportive public policies are required to not only support the necessary transition but to ensure it occurs as rapidly as possible. The uncertainty regarding scale boundaries and the mounting evidence that irrevocable harm could occur in a matter of decades, suggests haste would be prudent. During the transition phase from the current emphasis on material throughput to a sustainable economy, principles of social justice give priority to the world's poorest peoples to be the primary recipients of continued economic growth. This is perhaps one of the most difficult of issues for decision makers in wealthier countries. They fear that even a slow down in economic growth would be the beginning of their political demise. They appear to be operating on the assumption that electoral happiness is dependent on continued economic growth and ever increasing material well being. Considerable evidence suggests otherwise.

---

[66] The Factor Four concept visualizes a quadruple increase in resource efficiency using existing methodologies whilst avoiding negative impacts on the overall quality of life. The concept aims for society to last twice as long or enjoy twice as much whilst using half the resources and placing half the pressure on the environment. http://www.gdrc.org/sustdev/concepts/12-f4.html

## 6. About reduction

*Skeptic*: Economic growth has improved the human condition considerably over the last century and a half. It has allowed science and technology to improve the lives of hundreds of millions of people. To stop it now would be to turn the clock back and prevent further progress.

*Response*: There is no denying that economic growth has contributed to the material well being of many. Accepting the sustainable scale argument does not mean denying these realities. But it does mean looking at all the evidence. The enormous wealth accumulated by those in the top 10% of the global population occurred by impoverishing millions more. While many people benefit from economic growth, the vast majority are disadvantaged by the same processes. More relevant from a sustainable scale perspective is that the accumulation of this wealth is inextricably linked to the disruption of every major ecosystem and biogeochemical system on the planet. Continuing to accumulate financial wealth using the existing economic model will only contribute to the further disruption and possible destruction of these same global systems. Continued economic growth involving ever increasing material throughput is a biophysical impossibility. Even if no one had been harmed by the dominant economic model from a justice perspective, the potential harm to ecosystem functioning and human civilization means attention to the sustainable scale issue is urgently needed. Dealing with sustainable scale issues would actually represent progress, not a retreat. Addressing sustainable scale would represent progress in terms of using or devising scientific knowledge and technologies, as well as social institutions that ensure the sustainable use of naturally occurring ecosystem services. Addressing sustainable scale would also represent progress in terms of recognizing there are fatal flaws in our dominant mythologies so that individuals and nations can take control of their futures and ensure their survival. Any contest between immutable laws of nature and a defunct economic theory will not be won by relying on what produced benefits for the relatively few in the past. In thinking about solutions to the sustainable scale challenge it is useful to make a distinction between growth and development. Growth refers to a physical or quantitative increase in size; this is the meaning that has been used consistently throughout this discourse. It is economic growth based on ever increasing material throughput that is pushing us beyond sustainable scale. By way of contrast, development refers to a qualitative improvement. An economy based on qualitative improvement, rather than material throughput, would help address the challenge of sustainable scale. Such an approach can be considered in terms of moving from an economy based on economic growth to one based on economic development. Such a direction indicates that broadly based progress will be a very integral part of implementing sustainable scale relevant policies.

## 7. About technology

*Skeptic*: Granted that environmental problems are real and significant, we must keep in mind that technology has saved us before and will likely save us again. We are likely to devise new source of clean energy and other technologies that will allow us to easily avoid exceeding sustainable scale.

*Response*: Technology will continue to play an important role in dealing with sustainable scale issues. Currently, technologies are available to radically increase resource productivity by 400% or more, and new technologies are under development that promise increases of 1000% or more. There are several points to make about the role of technology in achieving sustainable scale:

- despite the availability of technologies that can increase resource productivity by 400% or more and simple energy conservation measures that are available across a broad range of applications, this knowledge is not being used. Few organizations require application of the most resource productive technologies available; no government regulation makes such a demand. Useful technology is available; the political will to use it is not.

- if all existing technologies that are environmentally friendly were the only ones in use, we would at least have more time before coming up against sustainable scale boundaries. But we are far from this ideal and even a transition to resource productive technologies will take time. Without knowing where the sustainable scale boundaries are, we have no idea how much time we have for a transition that has hardly begun.

- the economic model that is driving material throughput is itself a complex human system, consisting of many institutions which are mutually supportive. Specific technological innovations, regardless of how much they might contribute to alleviating the sustainable scale problem, will not be adopted if they do not fit into this system. Policies are needed that require such technologies to be explored and adopted. It is considerably easier to modify economic mechanisms to adapt to these new technologies than it is to change the laws of nature.

- an economy based on continually increasing material throughput will eventually and inevitably come up against sustainable scale boundaries. Using environmentally friendly technologies as soon and as extensively as possible is highly desirable. This would be a useful but inadequate step in dealing with the challenge of sustainable scale. Sustainable scale needs to be dealt with directly.

- relying on innovative technologies to solve the sustainable scale problem is a high risk strategy. New technologies first have to be invented, will take time to be implemented, and often have unintended and negative impacts that take years to identify. Indeed, the history of environmental damage involves the discovery of significant harm occurring from what originally appeared to be benign and useful technologies (e.g. DDT, CFCs, industrial agriculture, nuclear energy, fossil fuels, etc). To have faith in such a process to avoid the issue of sustainable scale, when we are not even using the clean technologies we have, seems unwarranted. We should certainly be exploring new technologies, but these should not be the main strategy for dealing with the problem of sustainable scale.
- it is also useful to keep the entropy law in mind when considering technological innovations. Technology only transforms energy from a more organized and ordered state to a less organized and ordered state – i.e. it inevitably (according to the entropy law) degrades whatever materials are involved in its processes.

## 8. About ingenuity

*Skeptic*: Human history has many examples of substituting new resources and sources of energy for old ones that were running out. Human ingenuity and creativity is infinite and will be used to solve the challenge of sustainable scale.

*Response*: Human history is indeed a story of new resources and sources of energy. And each change contributed to more economic activity and financial wealth.

There are several points to make about this history:

- as the world moved from wood to coal to petroleum as sources of energy, more energy became available, more economic growth was stimulated and more general wealth generated. The same transition also resulted in considerably more pollution and ecosystem disruption. This is an inevitable consequence of the entropy law. More economic activity means more degradation of matter and energy, and the faster the technologies allow us to proceed, the faster the degradation.

- the same is true of the broader range of resources that have been appropriated for human use. We now mine a considerable range of non-renewable resources that were not even known 200 years ago. And our technologies involve new combinations of these basic minerals and metals that never existed before through the wonders of modern chemistry. The result is increasing amounts of toxic substances entering global ecosystems. The process of extracting these resources from ever poorer sources (in terms of yield per ton) is also increasing ecosystem disruption on a massive scale. There are costs to human ingenuity that are not always taken into account.

- the history of new resources is also the history of using materials of increasingly low entropy (ie. naturally occurring resources of great value because of the available energy they represented). Petroleum is a more concentrated form of energy than coal, and coal more than wood. Iron ore and nickel are of lower entropy than clay and mud. There is no question that access to these high energy sources and rarer minerals and metals has contributed significantly to economic growth and increased material well-being for millions. But to fully appreciate this history, the negative aspects of this transition must also be recognized. The use of fossil fuels is threatening the stability of the global carbon cycle which regulates global climate, and the impact of toxic materials is threatening ecosystems around the world.

167

- at some point in human history we will run out of valuable low entropy materials that are nonrenewable. The faster they enter the economic cycle, the sooner will they be depleted and humanity will be confined once again to using clay and mud. There is a limit to the substitution of one resource for another in a finite biophysical world.
- the human ingenuity perspective is similar to the *"technology will save us"* view.
- there is no question that ingenuity and creativity are uniquely human traits. But unless they are applied to the scale issue, they cannot contribute to a solution. The application of these traits to particular, novel areas does not seem to occur until some catastrophe has drawn attention. Once a scale boundary has been exceeded, no amount of ingenuity will recreate an ecosystem that took millions of years, along with a complex and unique set of circumstances to evolve. For scale issues, prevention is the only cure.
- the complexity, pervasiveness, seriousness and unprecedented nature of the scale problem, along with the irrevocable loss if we exceed scale boundaries, suggests that the most creative minds on the planet should be encouraged to stop the damage currently being done.

## 9. About market

*Skeptic*: If we were really running out of natural resources their prices would be increasing. The scarcer they become the higher their price should be. This is a basic fact of economics.

*Response*: Supply and demand dynamics are powerful market mechanisms and can do an excellent job of establishing prices in a competitive market. However, there are several aspects of economic theory and practice that significantly distort the prices of most if not all natural resources:

- economic theory treats natural resources as income rather than as a draw down of natural capital. Accounting for natural resources follows different rules than accounting for other matters. The result is lower natural resource prices.

- natural resources only have monetary value once they are extracted. Consequently the size of the above ground reserve influences marginal costs – not the amount left in the earth. Scarcity is not viewed from the perspective of what is left in the ground; market prices are therefore disconnected from actual reserves in the ground.

- large financial government subsidies characterize mining and petroleum extraction. Hundreds of billion of dollars are spent annually to keep market prices of these commodities low. Additional subsidies in such areas as transportation also lower their final price.

- political considerations are also a significant factor in the setting of many natural resources. Oil, for example, is keep artificially low because of the potentially negative impact price rises could have on the global economy. The central role that energy plays in the global economy means that these lower energy prices also make other prices lower than they would be under open market pressures. Lower prices mean more use of materials and therefore more material degradation.

- political and economic pressures can also be brought to bear on many developing nations to keep natural resource prices low. In several smaller developing countries, resource extraction may be the major or sole source of foreign revenue, and/or opportunities for great wealth. If appropriate democratic institutions are not in place, the controlling ruling elite is often only too willing to make deals with foreign investors as personal wealth enhancement strategies.

- another major factor in the distorted pricing of most natural resources has to do with the fact that their extraction and use often result in considerable costs to other parties. These costs are rarely calculated in the market price. The costs of displacement of indigenous peoples and their consequent loss of culture and livelihood as a result of mining do not generally enter pricing. Nor do the costs of smog and air pollution that result in health problems; nor the costs of habitat destruction that accelerates the extinction of species.

If all of these real costs directly associated with natural resource use were actually included in the price, less of these resources would be used. It is not in the short term financial interests of industries involved in natural resource extraction to have these costs reflected in prices. Economic practices are therefore permitted by public policies which allow these costs to be externalized (borne by others than the industry involved and those who use its products). Where attempts have been made to calculate these externalized costs, they are often many times higher than the market price of the commodity:

- because resource prices are so distorted and considerably lower than they would otherwise be if market forces were left to operate, resources are overused and wasted. This phenomenon has implications not only for current impacts on ecosystem functioning, but also on the resources available to future generations.

- because various economic practices serve to keep resource prices low, prices are not a good indication of resource scarcity. There is considerable other evidence that natural resources are indeed becoming scarcer. World oil production is expected to peak sometime in the future, even though there are more oil wells pumping more oil now than ever before. Production will begin to decline because the new reserves are not being discovered as quickly as known reserves are being depleted. This is occurring despite ever increasingly sophisticated techniques for detecting oil deposits.

Further evidence of decreasing natural resources comes from the declining yield produced by mining ores of various kinds. A ton of nickel ore, for example, now yields only one $1/100^{th}$ of the yield when nickel was first mined. Nonrenewable resources that are becoming scarcer are in fact part of the scale problem. But the far more significant issue is the even greater scarcity of renewable resources. Several fish stocks have been depleted and thousands of species have been made extinct by various kinds of economic activity. In addition, the disruption of ecosystem services by economic activities is threatening the planet's life support systems. This is at the heart of the scale problem.

## 10. About the time

*Skeptic*: Even if the scale issue is real, we are likely still a very long way from any scale boundaries. If we were close to catastrophe surely scientists and politicians around the world would be active to fix the problem. Things seem to be getting better rather than worse and human ingenuity is such that we are likely to be able to solve the problem when it arises.

*Response*: There is no question that the issue of sustainable scale is real. One does not have to refer to basic laws of science to know that infinite growth in a finite world is an impossibility. The real question is how close or far away are we from a sustainable scale boundary that, if exceeded, would mean global catastrophe. The precise answer to this question is that we do not know. But this is hardly a reassuring answer, given the magnitude and irrevocable nature of a major ecosystem crash. Scientists have indeed been sending out warning signals on many issues related to scale. Evidence and logic indicate that some sustainable scale boundaries have already been breached, and others may occur in decades rather than centuries. But sustainable scale is not on the political agenda, and there are many factors working against giving it serious consideration.

## 11. About alternatives

*Skeptic*: Even if the scale problem is real, there must be another way to solve it other than giving up economic growth. We have gained too much from such growth and it would be too difficult to give up the material well being that growth has produced.

*Response*: The basic laws of science and common sense tell us that sustainable scale is a real problem. And enough scientists are sending out warning signals to suggest we should all be concerned. This is the first and most important step: to seriously consider the relevance and immediacy of the dangers posed by exceeding sustainable scale, at a national and international level. If alternative analyses of the problem identify causes other than ever increasing amounts of material throughput, then the appropriate adjustments can be made in terms of policy changes to ensure we do not exceed sustainable scale. The important point is to ask the significant questions about the sustainable scale of our global economy and address the findings. It is a misconception, however, to equate dealing with sustainable scale with a reduced quality of life, even if the cause is economic growth based on material throughput. Considerable evidence from around the world indicates that financial wealth and material goods are not the most important source of human happiness and well being. These issues are important, but only below a certain level (*determinants of human happiness and wellbeing*[67]). Once a comfortable sufficiency is reached, in terms of food, shelter, basic amenities, education and healthcare, more money does not contribute to more happiness or well being. Beyond comfortable sufficiency human happiness is primarily determined by a person's connection to friends, family and community – all non-market factors. Dealing with sustainable scale boundaries may mean giving up our current rate of consuming material goods. This does not mean giving up human happiness or well being. But it does mean reorienting social priorities on a global level and using human ingenuity in a manner very different from current applications. Indeed, a greater focus on these non-market factors, which are wholly compatible with sustainable scale, could contribute to greater human happiness and well being.

## 12. About decisions

*Skeptic*: If resources and ecosystem services are indeed limited and we are approaching these limits, then our government must do everything in its power to ensure that we continue our way of life, and others will simply have to adjust. This is not a just world and we should use any means available to us to ensure we survive and are well off. Survival of the fittest will be best for everyone.

*Response*: There is no doubt that ecosystems are being challenged, perhaps beyond their limits. The idea that any nation could appropriate the remaining resources and ecosystem services for their own use is both morally repugnant and realistically impractical.

Any national policy based on an *"us first"* approach would violate moral precepts at the heart of each of the world's major spiritual

---

[67] In brief, the determinants of wellbeing have into four key dimensions of life: political, economic, social, and personal. In general, the participants believed that they influenced wellbeing through their interactions with each other.

traditions. All notions of justice and respect for the legitimate rights of others would have to be violated; people and nations that resisted would have to be physically subdued and resources would have to be stolen. Theft and murder of such magnitude, regardless of the rationalizations that would be concocted, would be a transgression of basic moral precepts that have guided humanity for centuries. There is no doubt that there would be religious leaders within the *"us first"* nation who would justify such a policy. It would be interesting to see if any religious leaders outside the *"us first"* nation were to agree. Citizen in the *"us first"* nation would also have to be convinced that such an approach is not only moral, but also doable. An *"us first"* policy would involve unlawful exploitation of global resources, denying the rights of others, and would undoubtedly require extensive military actions to secure access to these resources. An extensive global reach would be required unless the *"us first"* nation was self sufficient - an unlikely circumstance given the current advance of economic globalization. In either case, the majority of citizens in the *"us first"* nation would likely experience a significant reduction in their customary way of life because the same global reach would be impractical. Since other peoples' and nations' survival would be at stake, bloody conflicts would increase. As the nation that pursued this policy moved ahead, other nations would become increasingly impoverished, desperate and aggressive. Attacks on the *"us first"* nation itself, as well as its global supply chain, would increase. International cooperation on a wide range of issues would come to an end, as other nations either joined forces to thwart the *"us first"* nation, or to compete with it. Such a world would likely hasten the very ecosystem degradation the *"us first"* strategy was intended to delay. Life within the *"us first"* nation would change dramatically. In addition to not having access to the same range of previously available resources (because of unpredictably interrupted supply chains in a much more hostile world), attacks on the homeland itself would likely increase. The kinds of protective actions the *"us first"* nation would enact would severely restrict citizens freedoms and civil rights. The decline in quality of life that the policy was designed to avoid would be hastened. Even if the *"us first"* policy allowed some semblance of continued prosperity, it would be a pyrrhic victory at best. It would simply delay the destruction that is inevitable from breaching sustainable scale. In such circumstances, having children would take on a new meaning, highlighting the hopeless future that would lie ahead. Slowly sliding into a world where sustainable scale could be breached is a frightening enough scenario. A world where one or more nations openly or secretly embark on a policy of unbridled self interest is even worse, rejecting and thrashing the very best that humans are capable of. Survival for all is most hopeful by building on these very virtues and values that an *"us first"* policy would violate. Malthus was indeed wrong in his timing about the relationship between food production and supply, and population growth, for he failed to anticipate industrialized agriculture. But the essential connection between food supply and population is one which has characterized the rise and fall of societies around the world. Today's population is poised to increase substantially, at a time when there are greater and greater losses of cropland and soil fertility. The issues he raised are still very much with us, and need urgent attention now more than ever.

# A Synthesis

Now, a quick review about what we about what we learned from the previous pages.

This part:

- provides an overview of the concept of sustainable scale (*Understanding Scale*).
- identifies why it is vitally important for human well being (*Importance of Scale*).
- outlines common causes of scale problems (*Causes of Scale Problems*).
- describes existing global scale problems (*Areas of Concern, we'll go deeply in the issue in Volume II*)).
- and identifies *Attractive Solutions(we'll go deeply in the issue in Volume III)*.

Existing global scale problems are reviewed in terms of their current status, the adequacy of current steps being taken to deal with them, and what else is needed to achieve and maintain sustainable scale.

The dialogue with skeptics provided other occasions to reflect, especially for those who either challenge the idea that sustainable scale is a serious problem or who challenge the proposed solutions.

*Lessons*

The basic lessons from this review include:

- scale is about appropriate size - whether the size of one thing is appropriate relative to another.
- sustainable scale is about the physical size of global economic activities relative to the biophysical limits of the ecosystems which contain and sustain them.

- a review of scientific evidence indicates there are currently many areas of economic activity in which sustainable scale is being exceeded - i.e. the throughput is overwhelming the ecosystems' capacities to regenerate the sources and sinks upon which the economic activities depend.
- in addition, the production and consumption cycles associated with these specific levels of throughput are also degrading the ecosystems' capacities to continue providing a variety of critical life supporting services.
- in aggregate, global economic activities are overshooting the living planet's ability to support human civilization as it currently exists.

*Overarching Goals*

From the perspective of human well being and happiness, the current economic paradigm and its consequences are both ecologically unsustainable, and morally unjust. Limits to material throughput are required for ecological sustainability (*A Sustainable Scale Perspective, in Vol. III*). Redistribution of the wealth generated by ecosystem services is required for social justice. It is unlikely that ecological sustainability is possible without social justice. The concept of Optimal Scale captures both goals and is identified as the policy priority to achieve sustainable scale (see *Sustainable Scale*, and *A Sustainable Scale Perspective, in Vol. III*).

*Actions Needed*

A wide variety of approaches are available to achieve sustainable scale; these include:
- clear policy goals as articulated by Optimal Scale (*A Sustainable Scale Perspective, in Vol. III*).
- clear visions for a sustainable and just future (*Visions For A Sustainable Future, in Vol. III*).

- clearly *Understanding Human Happiness and Well Being*, in *Vol. III*.
- public policies that integrate a sustainable scale perspective (*Supportive Public Policies*, in *Vol. III*).
- population policies compatible with Optimal Scale (*Population*, in *Vol. II*).
- economics for community (*Economics For Community*, in *Vol. III*).
- sustainable business practices (*Sustainable Business Practices*, in *Vol. III*).
- institutions to implement sustainable scale policies (*Institutions for a Sustainable Future*, in *Vol. III*).
- lifestyles that optimize human well being and happiness within the biophysical limits of ecosystems (*Lifestyle Solutions*, in *Vol. III*).

It is clear that there are a variety of *Attractive Solutions* to dealing with the existing sustainable scale problems we face. It is also clear that if these issues are not resolved soon, it will become increasingly difficult and costly to deal with them in the future. Unresolved, they will eventually lead to the collapse of human civilization as we know it. Still, it's clear that the high levels of material throughput which characterize developed countries are not necessary for human well being and happiness - they are both wasteful and harmful. The overriding obstacle to achieving sustainable scale is political will. While many technical and social challenges exist, there is considerable reason for optimism if the solutions available are implemented. An informed and supportive public is badly needed.

Ultimately, it is up to each of us to not only adjust our own lifestyles to be compatible with sustainable scale, but to lobby governments and decision makers at all levels and sectors to implement sustainable scale policies and practices. Our future depends on it.

# Acknowledgments
## And
## Heartfelt thank to

- *Brian Czech*, has a B.S. in wildlife ecology from the University of Wisconsin-Madison, an M.S. in wildlife science from the University of Washington, and a Ph.D. in renewable natural resources from the University of Arizona. https://steadystate.org/brian-czech/
https://steadystate.org/

- *David Batker* directs the APEX Center for Applied Ecological Economics.
https://www.uvm.edu/gund/profiles/david-batker
- *Dr. Herman E. Daly* is currently Professor at the University of Maryland, School of Public Affairs. https://steadystate.org/herman-daly/
- *Dr. Joshua Farley* received his undergraduate degree in Biology from Grinnell College in 1985, his Master in International Affairs and a Certificate in Latin American and Iberian Studies from Columbia University's School for International and Public Affairs in 1990, and his Ph.D. in Agricultural, Resource and Managerial Economics from Cornell University in 1999. https://www.uvm.edu/rsenr/profiles/joshua_farley
https://fieldstudies.org/

..... the Advisory Panel of the Sustainable Scale Project to which we want to give new life with updates of new data from these first twenty years of the new millennium.

# Selected Readings

**Income**

Hicks, J. *"Value and Capital"*, 2nd edition - Oxford: Claredon, 1948.

**Ecosystem Functions and Services**

Daily, Gretchen (ed.). *"Nature's Service"* - Washington: Island Press, 1997.

Daily, Gretchen, T. Soderqvist, S. Aniyar, K. Arrow, P. Dasgupta, P. Ehrlich, C. Folke, A. Jansson, B. Jansson, N. Kautsky, S. Levin, J. Lubchenco, K. Maler, D. Simpson, D. Starrett, D. Tilman and B. Walker. "The Value of Nature and the Nature of Value" - *Science Magazine*. V.289.5478 (21 July 2000): 395-396.

Meffe, Gary, L. Nielsen, R. Knight and D. Schenborn. *"Ecosystem Management: Adaptive Community-Based Conservation"* - Washington: Island Press, 2002.

**Natural Capital and Income**

Costanza, Robert and H. Daly. *"Natural Capital and Sustainable Development"* - Conservation Biology, 6.1 (March 1992): 37-46.

Ekins, Paul, S. Simon, L. Deutsch, C. Folke and R. de Groot. *"A framework for the practical application of the concepts of critical natural capital and strong sustainability"* - Ecological Economics 44 (2003): 165–185.

Ekins, Paul. *"Identifying critical natural capital: Conclusions about critical natural capital"* - Ecological Economics 44 (2003): 277-292.

Ekins, Paul, C. Folke and R. de Groot. *"Identifying critical natural capital"* - Ecological Economics 44 (2003): 165-185.

**Thermodynamics**

Georgescu-Roegen, Nicholas. *"The Entropy Law and the Economic Process"* - Cambridge, MA: Harvard University Press, 1971.

Rifkin, J. and Ted Howard. *"Entropy"* - Toronto: Bantam Books, 1980.

**Carrying Capacity**

Daily, G. and P. R. Ehrlich. *"Population, sustainability, and Earth's carrying capacity"* - BioScience 42 (1992): 761-771.

**IPAT**

Waggoner, P. E. and J. Ausubel, *"A Framework for Sustainability Science: A Renovated IPAT Identity"* - Proceedings of the National Academy of Sciences 99.12 (2002): 7860.

Kates, Robert W. *"Population and Consumption: What We Know and What We Need to Know"* - Environment 42.3 (2000): 10.

**Limits to Growth**

Meadows, Donella, J. Randers and D. Meadows. *"Limits to Growth"* - New York: Universe Books, 1972.

Meadows, Donella, J. Randers and D. Meadows. *"Beyond the Limits"* - White River Junction, VT: Chelsea Green Publishing Co., 1992.

Meadows, Donella, J. Randers and D. Meadows. *"Limits to Growth: The Thirty Year Update"* - White River Junction, VT: Chelsea Green Publishing Co., 2004.

**Critical Natural Capital**

Balmford, Andrew, A. Bruner, P. Cooper, R. Costanza, S. Farber, R. Green, M. Jenkins, P. Jefferiss, V. Jessamy, J. Madden, K. Munro, N. Myers, S. Naeem, J. Paavola, M. Rayment, S. Rosendo, J. Roughgarden, K. Trumper and R. K. Turner. *"Economic Reasons for conserving wild nature"* - Science 297.5583 (9 August 2002): 950-953.

Baskin, Y. *"The Work of Nature: How the Diversity of Life Sustains Us"* - Washington: Island Press, 1997.

Chiesura, Anna and R. de Groot. *"Critical natural capital: a socio-cultural perspective"* - Ecological Economics 44 (2003): 219-231.

Deutsch, Lisa, C. Folke and K. Skanberg. *"The critical natural capital of ecosystem performance as insurance for human well-being"* - Ecological Economics 44 (2003): 205–217.

Ekins, Paul, S. Simon, L. Deutsch, C. Folke and R. de Groot. *"A framework for the practical application of the concepts of critical natural capital and strong sustainability"* - Ecological Economics 44 (2003): 165–185.

Ekins, Paul, C. Folke and R. de Groot. *"Identifying critical natural capital"* - Ecological Economics 44 (2003): 165-185.

Ekins, Paul. *"Identifying critical natural capital: Conclusions about critical natural capital"* - Ecological Economics 44 (2003): 277-292.

**Ecological Footprint**

Borgstrom, G.. *"The Hungry Planet"* - New York: Macmillan, 1965.

Wackernagel, M. *"Tracking the Ecological Overshoot of the Human Economy"* - *Proceedings of the National Academy of Sciences* 99.14 (July 2002): 9266-9271.

Wackernagel, Mathis and W. Rees. *"Our Ecological Footprint"* - Gabriola Island, BC: New Society Publishers, 1996.

**Material Flow**

Matthews, Emily, C. Amann, S. Bringezu, M. Fischer-Kowalski, W. Huttler, R. Kleijn, Y. Moriguchi, C. Ottke, E. Rodenburg, D. Rogish, H. Schandl, H. Schutz, E. van der Voet and H. Weisz. *"The Weight of Nations: Material Outflows from Industrial Economics"* - Washington: World Resources Institute, 2000.

**Millenium Assessment**

Millennium Ecosystem Assessment. *"Ecosystems and Human Well-being"* - Washington: Island Press, 2003.

**Panarchy**

Fraser, E., Figge, F., and W. Mabee. *"A framework for assessing the vulnerability of food systems to future shocks"* - *Futures* (2004).

Fraser, E.. *"Social vulnerability and ecological fragility: building bridges between social and natural sciences using the Irish potato famine as a case study"* - *Conservation Ecology*, 7.2 (2003): 9.

Gunderson, Lance and C. S. Holding. *"Panarchy: Understanding Transformations in Human and Natural Systems"* - Washington: Island Press, 2002.

"Resilience Alliance" www.resalliance.org

**Brief Overview**

Costanza, Robert, O. Segura and J. Martinez Alier (eds.). *"Getting Down to Earth: Practical Applications of Ecological Economics"* - Washington: Island Press, 1996.

Daly, Herman and J. Farley. *"Ecological Economics: Principles and Applications"* - Washington: Island Press, 2004.

Daly, Herman and K. Townsend (eds.). *"Valuing the Earth: Economics, Ecology, Ethics"* - Cambridge, MA: MIT Press.

Daly, Herman (ed.). *"Ecological Economics and the Ecology of Economics"* - Northampton: Edward Elgar, 1999.

Daly, Herman and J. Cobb. *"For the Common Good: Redirecting the Economy toward Community, the Environment and a Sustainable Future"* - Boston: The Beacon Press, 1989.

Daly, Herman. *"Beyond Growth: The Economics of Sustainable Growth"* - Boston: Beacon Press, 1996.

Faber, Malte, Reiner Manstetten and John Proops. *"Ecological Economics: Concepts and Methods"* - Cheltenham, UK: Edward Elgar, 1998.

Krishnan, Rajarman, Jonathan Harris and Neva Goodwin (eds.). *"A Survey of Ecological Economics"* - Washington: Island Press, 1995.

McNeill, J. R. *"Something New Under the Sun"* - New York: W. W. Norton Co., 2000.

Ponting, Clive. *"A Green History of the World"* - New York: Penguin Books, 1991.

## Moral Approaches

Brown, Peter. *"The Commonwealth of Life"* - Montreal: Black Rose Books, 2001.

Gardner, Gary. *"Invoking the Spirit: Religion and Spirituality in the Quest for a Sustainable World"* - *Worldwatch Paper* (164) December 2002.

# INDEX

www.ingramcontent.com/pod-product-compliance
Lightning Source LLC
Chambersburg PA
CBHW030629220526
45463CB00004B/1468